MW01202226

THE 100+ SERIES™

MATH PRACTICE

Essential Practice for Advanced Math Topics

Carson-Dellosa Publishing LLC
Greensboro, North Carolina

Visit *carsondellosa.com* for correlations to Common Core, state, national, and Canadian provincial standards.

Carson-Dellosa Publishing LLC
PO Box 35665
Greensboro, NC 27425 USA
carsondellosa.com

ISBN 978-1-48380-081-3

04-224197784

Table of Contents

Introduction

What are the Common Core State Standards for Middle School Mathematics?

In grades 6–8, the standards are a shared set of expectations for the development of mathematical understanding in the areas of ratios and proportional relationships, the number system, expressions and equations, functions, geometry, and statistics and probability. These rigorous standards encourage students to justify their thinking. They reflect the knowledge that is necessary for success in college and beyond.

Students who master the Common Core standards in mathematics as they advance in school will exhibit the following capabilities:

1. Make sense of problems and persevere in solving them.

Proficient students can explain the meaning of a problem and try different strategies to find a solution. Students check their answers and ask, "Does this make sense?"

2. Reason abstractly and quantitatively.

Proficient students are able to move back and forth smoothly between working with abstract symbols and thinking about real-world quantities that symbols represent.

3. Construct viable arguments and critique the reasoning of others.

Proficient students analyze problems by breaking them into stages and deciding whether each step is logical. They justify solutions using examples and solid arguments.

4. Model with mathematics.

Proficient students use diagrams, graphs, and formulas to model complex, real-world problems. They consider whether their results make sense and adjust their models as needed.

5. Use appropriate tools strategically.

Proficient students use tools such as models, protractors, and calculators appropriately. They use technological resources such as Web sites, software, and graphing calculators to explore and deepen their understanding of concepts.

6. Attend to precision.

Proficient students demonstrate clear and logical thinking. They choose appropriate units of measurement, use symbols correctly, and label graphs carefully. They calculate with accuracy and efficiency.

7. Look for and make use of structure.

Proficient students look closely to find patterns and structures. They can also step back to get the big picture. They think about complicated problems as single objects or break them into parts.

8. Look for and express regularity in repeated reasoning.

Proficient students notice when calculations are repeated and look for alternate methods and shortcuts. They maintain oversight of the process while attending to the details. They continually evaluate their results.

How to Use This Book

In this book, you will find a collection of 100+ reproducible practice pages to help students review, reinforce, and enhance Common Core mathematics skills. Use the chart provided on the next page to identify practice activities that meet the standards for learners at different levels of proficiency in your classroom.

Common Core State Standards* Alignment: Middle School Math Practice

Domain: Ratios and Proportional Relationships		Domain: Expressions and Equations	
Standard	**Aligned Practice Pages**	**Standard**	**Aligned Practice Pages**
6.RP.A.1	7, 20–22, 107, 108	6.EE.A.1	81, 82
6.RP.A.2	9–11	6.EE.A.2a	85
6.RP.A.3a	12, 25, 26	6.EE.A.2c	83–87
6.RP.A.3b	10, 11	6.EE.B.5	88
6.RP.A.3c	13–24	6.EE.B.7	86
6.RP.A.3d	25, 26	7.EE.B.3	89–91
7.RP.A.2a	8, 12	**Domain: Geometry**	
7.RP.A.2c	27–29	**Standard**	**Aligned Practice Pages**
7.RP.A.3	30–35	6.G.A.1	81, 92–95
Domain: The Number System		6.G.A.2	98, 99
Standard	**Aligned Practice Pages**	7.G.B.4	100, 101
6.NS.A.1	36–39	7.G.B.5	102
6.NS.B.2	40–45	7.G.B.6	92–99
6.NS.B.3	46–59	**Domain: Statistics and Probability**	
6.NS.B.4	60–64	**Standard**	**Aligned Practice Pages**
6.NS.C.5	65	6.SP.B.4	104
6.NS.C.6a	65, 66	6.SP.B.5a	103
6.NS.C.6c	66	6.SP.B.5b	103
6.NS.C.7a	67	6.SP.B.5c	105–107
6.NS.C.7b	67	7.SP.B.4	105–107
6.NS.C.7c	68	7.SP.C.5	109, 110
7.NS.A.1a	72	7.SP.C.6	103, 104
7.NS.A.1c	73	7.SP.C.8a	109, 110
7.NS.A.1d	68–75	7.SP.C.8b	108–110
7.NS.A.2a	76		
7.NS.A.2b	77		
7.NS.A.2c	70, 76–78		
7.NS.A.2d	8, 79, 80		

* © Copyright 2010. National Governors Association Center for Best Practices and Council of Chief State School Officers. All rights reserved.

Showing Ratios

Draw a model of each ratio.

1. The ratio of striped socks to solid-colored socks is 2:6.

2. The ratio of total dots to green dots is 9 to 2.

3. The ratio of candy bars to suckers is 10/1.

4. The ratio of yellow pencils to total pencils is 5:6.

There are three ways to write ratios: 5 to 6, 5/6, or 5:6. Write each ratio two other ways.

5. 2 to 3 _____ _____

6. 9 to 6 _____ _____

7. _____ 7/3 _____

8. _____ 5/9 _____

9. _____ _____ 3:6

10. _____ _____ 8:4

11. 12 to 9 _____ _____

12. 1 to 7 _____ _____

13. _____ _____ 8:3

14. _____ 12/5 _____

7.RP.A.2a, 7.NS.A.2d

Equality of Ratios

Decide if the two ratios are equal. If you check the "equal" box, finish the rest of the chart by writing the ratios in lowest terms three different ways. Follow the example.

	Ratios		Are ratios equal?	3 ways to write ratios		
1.	$\frac{12}{28}$	$\frac{21}{49}$	✔	$\frac{3}{7}$	3 : 7	$0.\overline{428571}$
2.	$\frac{49}{70}$	$\frac{35}{50}$				
3.	$\frac{6}{24}$	$\frac{5}{20}$				
4.	$\frac{45}{55}$	$\frac{18}{22}$				
5.	$\frac{2}{12}$	$\frac{9}{54}$				
6.	$\frac{85}{136}$	$\frac{15}{24}$				
7.	$\frac{12}{45}$	$\frac{28}{105}$				
8.	$\frac{18}{84}$	$\frac{22}{98}$				
9.	$\frac{85}{125}$	$\frac{51}{75}$				
10.	$\frac{42}{63}$	$\frac{30}{45}$				
11.	$\frac{39}{59}$	$\frac{52}{120}$				
12.	$\frac{52}{80}$	$\frac{78}{120}$				
13.	$\frac{36}{45}$	$\frac{28}{35}$				
14.	$\frac{21}{48}$	$\frac{49}{112}$				
15.	$\frac{32}{72}$	$\frac{12}{27}$				

Find the Rate

Find the rate for each item. Do your work in the space at the bottom of the page.

1. per hour if you drive 220 miles in 4 hours. _____
2. per minute if your heart beats 234 times in 3 minutes. _____
3. per can if you pay $1.68 for 6 cans of pop. _____
4. per hour if it snows 3 inches in 4 hours. _____
5. per minute if you type 240 words in 5 minutes. _____
6. per ounce if you spend $3.49 for a 20-ounce box of cereal. _____
7. per pound if you spend $1.30 for a 5-pound bag of potatoes. _____
8. per day if you consume 13,300 calories in 7 days. _____
9. per gallon if you pay $2.25 for a 0.5-gallon container of frozen yogurt. _____
10. per gallon if you pay $10.80 for 8 gallons of cider. _____

Rates

What is the unit rate? Do your work in the space at the bottom of the page.

1. 84 m in 6h_____

2. $92 for 23 books_____

3. 4500 km in 9 days _____

4. 198 in. in 9 days _____

5. $294 for 7 dresses _____

6. 36 dogs for 12 owners_____

7. 68 m in 4 s _____

8. 418 km in 38 days _____

9. 40 in. in 8 h_____

10. $1,200 for 4 plane tickets _____

11. 171 cm in 19 s _____

12. 468 m in 18 h _____

13. 63 m in 3 min _____

14. 90 toys for 6 babies_____

15. 12 exams in 6 days _____

16. 640 m in 80 days_____

17. $78 for 13 sandwiches _____

18. 98 km in 7 h_____

19. 64 travelers for 16 cars_____

20. $625 for 5 coats _____

More Rates

What is the unit rate? Do your work in the space at the bottom of the page.

1. 175 m in 7 h _____

2. 104 notebooks for 26 students _____

3. 48 km in 4 d _____

4. 162 km in 9 h_____

5. 112 candy bars for 14 students _____

6. $78 for 6 CD's _____

7. 159 in. in 53 s _____

8. 84 km in 12 h_____

9. 132 ft in 6 min_____

10. 288 marbles for 48 children _____

11. $38 for 19 hot dogs _____

12. 48 in. in 16 s _____

13. 72 km in 8 h _____

14. $49 for 7 pizzas_____

15. 66 m in 6 days _____

16. 361 km in 19 h_____

17. $26,000 for 2 cars _____

18. 21 m in 3 s_____

19. 42 classes in 7 days _____

20. $1,360 for 4 TV sets _____

Equivalent Fractions

Write yes or no under each ratio to tell whether or not it is equal.

1. $\frac{21}{9} = \frac{7}{3}$

2. $\frac{2}{4} = \frac{5}{9}$

3. $\frac{40}{50} = \frac{50}{40}$

4. $\frac{8}{4} = \frac{210}{105}$

Circle the equivalent fractions.

5. $\frac{12}{15} = \frac{28}{35}$

6. $\frac{63}{72} = \frac{28}{32}$

7. $\frac{3}{18} = \frac{10}{60}$

8. $\frac{16}{19} = \frac{17}{23}$

9. $\frac{18}{30} = \frac{24}{45}$

10. $\frac{24}{34} = \frac{60}{85}$

11. $\frac{60}{70} = \frac{8}{9}$

12. $\frac{12}{27} = \frac{16}{36}$

13. $\frac{30}{42} = \frac{20}{28}$

14. $\frac{47}{52} = \frac{33}{38}$

15. $\frac{12}{33} = \frac{28}{77}$

16. $\frac{16}{24} = \frac{6}{9}$

17. $\frac{36}{96} = \frac{27}{72}$

18. $\frac{35}{50} = \frac{14}{20}$

19. $\frac{9}{11} = \frac{72}{99}$

20. $\frac{35}{42} = \frac{20}{28}$

21. $\frac{22}{25} = \frac{17}{20}$

22. $\frac{18}{24} = \frac{6}{8}$

23. $\frac{6}{45} = \frac{14}{105}$

24. $\frac{36}{45} = \frac{42}{55}$

Percents

Find the percents.

1. 56% of 300 _____
2. 14% of 50 _____
3. 10% of 60 _____
4. 20% of 39 _____
5. 30% of 18 _____
6. 15% of 50 _____
7. 70% of 80 _____
8. 10% of 15 _____
9. 50% of 18 _____
10. 90% of 130 _____
11. 40% of 65 _____
12. 80% of 120 _____

Figure the percent.

13. Find 85% of 80. _____

14. 32 is what percent of 96? _____

15. What percent of 75 is 15? _____

16. Find 35% of 160. _____

17. Find 18% of 1,200. _____

18. Find 150% of 50. _____

19. 22 is what percent of 50? _____

20. What percent of 40 is 100? _____

21. 45% of what number is 9? _____

22. 510 is 85% of what number? _____

More Percents

1. How much is 125% of 72? _____

2. 220% of 8 is how much? _____

3. How much is 12% of 130? _____

4. 0.5% of 400 is how much? _____

5. How much is 60% of 30? _____

6. 70% of 90 is how much? _____

7. How much is 215% of 420? _____

8. 55% of 80 is how much? _____

9. How much is 4% of 16? _____

10. 16% of 85 is how much? _____

11. How much is 35% of 44? _____

12. 6% of 20 is how much? _____

13. How much is 2.5% of 140? _____

14. 22% of 20 is how much? _____

15. How much is 14.8% of 50? _____

16. 92% of 300 is how much? _____

17. How much is 105% of 25? _____

18. 35.8% of 190 is how much? _____

19. 9% of 200 is how much? _____

20. How much is 150% of 150? _____

What Percent?

Solve. Do your work in the space at the bottom of the page.

1. What percent of 20 is 8? _____

2. 50 is what percent of 16? _____

3. 18 is what percent of 120? _____

4. What percent of 200 is 121? _____

5. What percent of 4 is 5? _____

6. 3 is what percent of 250? _____

7. 15 is what percent of 80? _____

8. What percent of 24 is 33? _____

9. What percent of 25 is 29? _____

10. 39 is what percent of 75? _____

11. 28 is what percent of 40? _____

12. What percent of 160 is 55? _____

13. What percent of 200 is 111? _____

14. 120 is what percent of 600? _____

15. 193 is what percent of 80? _____

16. What percent of 24 is 21? _____

17. What percent of 250 is 193? _____

18. 36 is what percent of 96? _____

19. 18 is what percent of 48? _____

20. What percent of 220 is 44? _____

21. What percent of 8 is 20? _____

22. 8 is what percent of 128? _____

23. 15 is what percent of 125? _____

24. What percent of 160 is 88? _____

6.RP.A.3c

Finding the Total Number

1. 70% of what number is 56? _____

2. 76 is 95% of what number? _____

3. 9.24 is 14% of what number?_____

4. 120% of what number is 72? _____

5. 2% of what number is 1.92?_____

6. 9.2 is 8% of what number? _____

7. 288 is 144% of what number? _____

8. 92% of what number is 110.4? _____

9. 96% of what number is 144? _____

10. 79.2 is 198% of what number? _____

Find the number that equals X in each statement.

11. 35 is 12.5% of X. _____

12. 45% of X is 9. _____

13. 40% of X is 82. _____

14. 0.5% of X is 6. _____

15. 92% of X is 368. _____

16. 10.5 is 14% of X. _____

17. 18 is 20% of X. _____

18. 52 is 130% of X. _____

19. 65% of X is 104. _____

20. 12 is 15% of X. _____

Percent Increase and Decrease

Solve. Do your work in the space at the bottom of the page.

1. What is the percent increase from 12 to 60? _____

2. What is 70 decreased by 20%? _____

3. What is 66 increased by 88%? _____

4. What is the percent decrease from 300 to 12? _____

5. What is the percent decrease from 72 to 45? _____

6. What is 90 decreased by 13%? _____

7. What is 480 decreased by 95%? _____

8. What is the percent decrease from 105 to 42? _____

9. What is the percent increase from 16 to 93? _____

10. What is 120 increased by 13%? _____

11. What is 175 decreased by 80%? _____

12. What is 60 increased by 28%? _____

13. What is the percent decrease from 220 to 33? _____

14. What is the percent increase from 40 to 51? _____

15. What is 94 increased by 30%? _____

16. What is 105 decreased by 15%? _____

17. What is the percent decrease from 168 to 63? _____

18. What is 90 increased by 95%? _____

19. What is 15 decreased by 4%? _____

20. What is the percent increase from 72 to 81? _____

Percents and Proportions

Write as a proportion and solve. Do your work in the space at the bottom of the page.

1. What number is 17% of 200? _____

2. 16 is what percent of 80? _____

3. 25% of what number is 35? _____

4. What percent of 90 is 27? _____

5. 112 is what percent of 400? _____

6. What number is 140% of 280? _____

7. 12% of what number is 66? _____

8. What percent of 30 is 54? _____

9. What number is 15% of 40? _____

10. What percent of 90 is 72? _____

11. 495 is what percent of 500? _____

12. 125% of what number is 85? _____

13. 70% of what number is 42? _____

14. What number is 20% of 95? _____

15. 216 is what percent of 600? _____

16. What number is 230% of 40? _____

17. What percent of 900 is 792? _____

18. 27 is what percent of 90? _____

19. What number is 175% of 220? _____

20. 18% of what number is 117? _____

21. 42 is what percent of 120? _____

22. What percent of 260 is 247? _____

23. What percent of 315 is 63? _____

24. 40% of what number is 16? _____

25. 16 is what percent of 50? _____

26. What number is 65% of 80? _____

Ratios as Percents

Write the following rates as percents.

1. $\dfrac{2}{5}$

2. $\dfrac{12}{25}$

3. 17 : 20

4. $\dfrac{11}{32}$

5. 23 : 250

6. 0.6875

7. $\dfrac{3}{5}$

8. 0.04

9. 7 : 16

10. 0.5625

11. 0.7

12. 0.525

13. 5 : 8

14. $\dfrac{17}{25}$

15. $\dfrac{7}{8}$

16. $\dfrac{333}{500}$

17. 33 : 40

18. $\dfrac{23}{50}$

19. $\dfrac{21}{80}$

20. 11 : 200

Decimal, Percent, Fraction

Write each decimal. Change each decimal to a percent. Change the percents to fractions. Write the fractions in lowest terms.

		Decimal	Percent	Fraction
1.	sixty hundredths	_____	_____	_____
2.	thirty-eight hundredths	_____	_____	_____
3.	fifteen hundredths	_____	_____	_____
4.	ninety hundredths	_____	_____	_____
5.	fifty-nine hundredths	_____	_____	_____
6.	twenty-one hundredths	_____	_____	_____

Write each decimal. Change each decimal to a percent. Change the percents to fractions. Write the fractions in lowest terms.

		Decimal	Percent	Fraction
1.	thirty-nine hundredths	_____	_____	_____
2.	two hundredths	_____	_____	_____
3.	fifty-one hundredths	_____	_____	_____
4.	eighty hundredths	_____	_____	_____
5.	twelve hundredths	_____	_____	_____
6.	eighty-three hundredths	_____	_____	_____

Write each decimal. Change each decimal to a percent. Change the percents to fractions. Write the fractions in lowest terms.

		Decimal	Percent	Fraction
13.	twenty-five hundredths	_____	_____	_____
14.	six hundredths	_____	_____	_____
15.	seventy-two hundredths	_____	_____	_____
16.	twelve hundredths	_____	_____	_____

Decimals and Fractions

Give the correct fraction for each decimal and decimal for each fraction.

1. $\frac{4}{10}$

2. $\frac{9}{100}$

3. 0.75

4. $\frac{1}{4}$

5. 0.6

6. $\frac{7}{14}$

7. .20

8. .09

9. $\frac{6}{8}$

10. 0.2

11. $\frac{6}{20}$

12. 0.83

Write each percent as a fraction or mixed number in simplest form and each fraction as a percent.

13. 96%

14. 175%

15. $\frac{1}{2}$

16. $33\frac{1}{3}\%$

17. $\frac{3}{8}$

18. $\frac{8}{100}$

19. 110%

20. $\frac{2}{5}$

21. $2\frac{1}{2}\%$

22. $1\frac{3}{100}$

23. $\frac{5}{8}$

24. 275%

Converting Numbers

Complete the chart by converting percents, decimals, and fractions.

	Percent	Fraction	Decimal
1.		$1\frac{7}{8}$	
2.			0.18
3.	2%		
4.		$\frac{17}{20}$	
5.			1.2
6.	135%		
7.			0.204
8.		$\frac{89}{200}$	
9.			0.33
10.	77%		
11.	40%		
12.		$\frac{17}{40}$	
13.	5.5%		
14.			1.95
15.		$3\frac{47}{50}$	
16.	264%		
17.			0.88
18.		$\frac{3}{10}$	
19.	12%		
20.			0.7175

	Percent	Fraction	Decimal
21.	19%		
22.		$\frac{7}{8}$	
23.		$1\frac{4}{5}$	
24.	128%		
25.		$\frac{1}{20}$	
26.			0.135
27.	86.5%		
28.			2.9
29.		$3\frac{13}{25}$	
30.			0.8125
31.	72%		
32.			1.44
33.		$\frac{61}{100}$	
34.			0.132
35.	27.6%		
36.		$\frac{151}{200}$	
37.			0.45
38.	87.5%		
39.		$\frac{111}{400}$	
40.			0.858

Rounding Percents

Write as a percent rounded to the nearest:

one of a percent

1. $\frac{4}{9}$

2. $\frac{9}{16}$

3. $\frac{3}{40}$

4. $\frac{27}{32}$

5. $\frac{17}{80}$

6. $\frac{2}{3}$

7. $\frac{7}{8}$

8. $\frac{17}{30}$

9. $\frac{8}{15}$

10. $\frac{8}{11}$

tenth of a percent

11. $\frac{8}{9}$

12. $\frac{22}{45}$

13. $\frac{4}{7}$

14. $\frac{5}{12}$

15. $\frac{77}{80}$

16. $\frac{13}{30}$

17. $\frac{111}{160}$

18. $\frac{5}{16}$

19. $\frac{19}{75}$

20. $\frac{17}{18}$

More Rounding Percents

Write as a percent rounded to the nearest:

hundredth of a percent

1. $\dfrac{7}{15}$　　2. $\dfrac{3}{11}$　　3. $\dfrac{5}{18}$　　4. $\dfrac{81}{160}$　　5. $\dfrac{6}{13}$

6. $\dfrac{21}{44}$　　7. $\dfrac{2}{9}$　　8. $\dfrac{97}{120}$　　9. $\dfrac{8}{21}$　　10. $\dfrac{6}{7}$

thousandth of a percent

11. $\dfrac{5}{19}$　　12. $\dfrac{13}{28}$　　13. $\dfrac{5}{26}$　　14. $\dfrac{13}{24}$　　15. $\dfrac{4}{21}$

16. $\dfrac{11}{12}$　　17. $\dfrac{2}{7}$　　18. $\dfrac{2}{13}$　　19. $\dfrac{1}{3}$　　20. $\dfrac{10}{11}$

Customary Measurements

Write the answer.

1. 5 feet 4 inches = _____ inches

2. 4 pounds = _____ ounces

3. 2 ½ gallons = _____ quarts

4. 80 ounces =_____ pounds

5. 90 minutes = _____ hours

6. 32 pints = _____ gallons

Write the equivalent for each measurement.

7. 1 yard = _____ inches

8. 1 1/2 yards = _____ inches

9. 2 feet = _____inches

10. 7 feet = _____ inches

11. 4 1/2 feet = _____ inches

12. 19 1/2 feet = _____ inches

13. 2 feet, 5 inches=_____ inches

14. 1 yard, 1 foot, 1 inch =__ inches

Write the equivalent measurement.

15. 4 gal 3 qt. = _____ qt.

16. 7 gal = _____ qt.

17. 72 fl oz. = _____ qt.

18. 2 qt. = _____ fl. oz.

19. 7 pt = _____ c

20. 4 pt = _____ fl. oz.

6.RP.A.3a, 6.RP.A.3d

Metric Measurements

Fill in the chart.

kilometer (km)	hectometer (hm)	dekameter (dam)	meter (m)	decimeter (dm)	centimeter (cm)	millimeter (mm)
1.	0.052					
2.			3.67			
3.				1.03		
4.		61				
5.						8,856
6.					32	
7. 0.73						
8.				406.9		
9.						3.8
10. 4.4						
11.			16			
12.	9.16					
13.					0.05	
14.		1.007				
15.		14.2				
16.				0.082		
17.					11.11	
18.			0.194			
19.	3					
20.						76.41
21. 90						
22.			7.03			
23.						118
24. 0.005						
25.					6.45	

Solving Proportions

Solve each proportion.

1. $\dfrac{8}{12} = \dfrac{6}{x}$

2. $\dfrac{3}{9} = \dfrac{x}{15}$

3. $\dfrac{12}{x} = \dfrac{8}{10}$

4. $\dfrac{15}{9} = \dfrac{x}{33}$

5. $\dfrac{9}{x} = \dfrac{30}{40}$

6. $\dfrac{7}{8} = \dfrac{14}{x}$

7. $\dfrac{x}{18} = \dfrac{4}{6}$

8. $\dfrac{6}{15} = \dfrac{x}{25}$

9. $\dfrac{6}{10} = \dfrac{60}{x}$

10. $\dfrac{9}{3} = \dfrac{12}{x}$

11. $\dfrac{x}{12} = \dfrac{12}{18}$

12. $\dfrac{3}{x} = \dfrac{10}{30}$

Solve each proportion.

13. $\dfrac{6}{8} = \dfrac{9}{x}$

14. $\dfrac{x}{3.6} = \dfrac{75}{100}$

15. $\dfrac{x}{4} = \dfrac{90}{15}$

16. $\dfrac{33}{x} = \dfrac{6}{2}$

17. $\dfrac{x}{5} = \dfrac{94}{1}$

18. $\dfrac{4}{28} = \dfrac{x}{42}$

19. $\dfrac{4}{x} = \dfrac{16}{20}$

20. $\dfrac{5}{10} = \dfrac{x}{25}$

21. $\dfrac{x}{7} = \dfrac{9}{21}$

22. $\dfrac{15}{45} = \dfrac{14}{x}$

23. $\dfrac{x}{90} = \dfrac{15}{60}$

24. $\dfrac{3.5}{10.5} = \dfrac{5}{x}$

Find x.

25. $\dfrac{8}{9} = \dfrac{9}{x}$

26. $\dfrac{x}{8} = \dfrac{10}{12}$

27. $\dfrac{4}{x} = \dfrac{8}{12}$

28. $\dfrac{36}{72} = \dfrac{7}{x}$

29. $\dfrac{x}{16} = \dfrac{3}{8}$

30. $\dfrac{20}{25} = \dfrac{7}{x}$

31. $\dfrac{7}{6} = \dfrac{9}{x}$

32. $\dfrac{9}{4} = \dfrac{10}{x}$

More Proportions

Solve the proportions. Do your work in the space at the bottom of the page.

1. $\dfrac{15}{n} = \dfrac{24}{128}$ 2. $\dfrac{n}{200} = \dfrac{28}{80}$ 3. $\dfrac{15}{110} = \dfrac{24}{n}$ 4. $\dfrac{4}{5} = \dfrac{n}{20}$

5. $\dfrac{54}{63} = \dfrac{n}{49}$ 6. $\dfrac{5}{n} = \dfrac{2}{38}$ 7. $\dfrac{n}{12} = \dfrac{45}{108}$ 8. $\dfrac{42}{n} = \dfrac{8}{12}$

9. $\dfrac{8}{30} = \dfrac{28}{n}$ 10. $\dfrac{n}{108} = \dfrac{20}{72}$ 11. $\dfrac{7}{13} = \dfrac{n}{78}$ 12. $\dfrac{56}{n} = \dfrac{7}{31}$

13. $\dfrac{n}{22} = \dfrac{63}{154}$ 14. $\dfrac{120}{n} = \dfrac{40}{55}$ 15. $\dfrac{95}{110} = \dfrac{n}{22}$ 16. $\dfrac{33}{44} = \dfrac{132}{n}$

17. $\dfrac{9}{27} = \dfrac{n}{21}$ 18. $\dfrac{10}{22} = \dfrac{30}{n}$ 19. $\dfrac{n}{120} = \dfrac{15}{100}$ 20. $\dfrac{68}{n} = \dfrac{102}{108}$

More Proportions

Solve the proportions. Do your work in the space at the bottom of the page.

1. $\dfrac{3}{n} = \dfrac{12}{76}$ 2. $\dfrac{15}{18} = \dfrac{n}{12}$ 3. $\dfrac{44}{77} = \dfrac{24}{n}$ 4. $\dfrac{n}{70} = \dfrac{36}{40}$

5. $\dfrac{84}{96} = \dfrac{35}{n}$ 6. $\dfrac{n}{27} = \dfrac{14}{63}$ 7. $\dfrac{32}{n} = \dfrac{28}{35}$ 8. $\dfrac{11}{12} = \dfrac{n}{60}$

9. $\dfrac{17}{20} = \dfrac{n}{120}$ 10. $\dfrac{30}{54} = \dfrac{20}{n}$ 11. $\dfrac{n}{24} = \dfrac{56}{64}$ 12. $\dfrac{11}{n} = \dfrac{44}{60}$

13. $\dfrac{18}{150} = \dfrac{15}{n}$ 14. $\dfrac{45}{72} = \dfrac{n}{56}$ 15. $\dfrac{77}{n} = \dfrac{42}{54}$ 16. $\dfrac{n}{56} = \dfrac{42}{49}$

17. $\dfrac{n}{41} = \dfrac{9}{123}$ 18. $\dfrac{35}{n} = \dfrac{5}{28}$ 19. $\dfrac{12}{52} = \dfrac{n}{39}$ 20. $\dfrac{40}{70} = \dfrac{32}{n}$

Sales Tax

Complete the chart. Round to the nearest cent.

Cost of Item	% Sales Tax	Tax Paid	Total Cost
1. $4.99	$3\frac{1}{4}$ %		
2. $12.50	5.65%		
3. $.58	$6\frac{3}{4}$ %		
4. $372.48	12%		
5. $111.20	$18\frac{1}{8}$ %		
6. $13.84	4.3%		
7. $25.25	7.11%		
8. $30.18	$8\frac{5}{8}$ %		
9. $441.89	9.0625%		
10. $580.60	14%		
11. $14.12	1.35%		
12. $8.19	6.8%		
13. $5.45	$5\frac{1}{4}$ %		
14. $613.20	22%		
15. $125.50	$11\frac{3}{8}$ %		
16. $220.16	$9\frac{1}{2}$ %		
17. $8.12	2.625%		
18. $9.00	8.9375%		
19. $16.85	19%		
20. $21.22	5.0375%		

Simple Interest

Complete the chart. Round to the nearest cent.

	Principal	Interest Rate Per Year	Time	Interest Earned
1.	$625.00	16%	6 months	
2.	$720.50	$7\frac{1}{2}$ %	1 year	
3.	$5,670.80	22%	9 months	
4.	$4,112.20	$11\frac{1}{8}$ %	$4\frac{1}{4}$ years	
5.	$905.60	14%	$5\frac{1}{2}$ years	
6.	$814.75	$5\frac{3}{4}$ %	4 years	
7.	$1,100.50	15%	3 months	
8.	$870.20	$8\frac{3}{8}$ %	$9\frac{3}{4}$ years	
9.	$415.15	$6\frac{1}{2}$ %	5 months	
10.	$6,540.50	11%	$1\frac{1}{4}$ years	
11.	$11,140.25	5.0375%	8 years	
12.	$26,500.75	8%	6 months	
13.	$408.50	2.625%	4 months	
14.	$910.80	21%	3 years	
15.	$12,540.00	$14\frac{5}{8}$ %	7 months	
16.	$9,750.50	12.0625%	$11\frac{1}{2}$ years	
17.	$810.40	$10\frac{1}{2}$ %	6 years	
18.	$4,480.10	4.6875%	3 months	
19.	$33,500.00	33%	$4\frac{1}{4}$ years	
20.	$18,549.99	9.6%	5 years	

Name_____

Interest

Complete the chart. Round to the nearest cent.

Principal	Interest Rate	Compounded	Time	Interest
1. $4,000.00	12.4%	semiannually	1 year	
2. $650.00	8%	quarterly	1 year	
3. $18,999.99	$7\frac{1}{4}$ %	annually	3 years	
4. $525.25	$19\frac{3}{4}$ %	monthly	2 months	
5. $27,428.20	21%	annually	2 years	
6. $5,000.00	$8\frac{3}{8}$ %	quarterly	1 year	
7. $16,888.75	14%	semiannually	18 months	
8. $21,050.25	10.6%	semiannually	2 years	
9. $9,420.55	16.2%	monthly	4 months	
10. $625.00	$18\frac{1}{8}$ %	monthly	2 months	
11. $718.99	20.5%	annually	2 years	
12. $330.20	17.9%	quarterly	6 months	
13. $890.15	$13\frac{7}{8}$ %	annually	2 years	
14. $10,000.00	8.85%	quarterly	1 year	
15. $15,980.00	9.8%	monthly	2 months	
16. $25,400.00	12.35%	annually	2 years	
17. $29,590.25	$19\frac{5}{8}$ %	semiannually	1 year	
18. $18,670.20	$21\frac{1}{4}$ %	quarterly	9 months	
19. $6,430.05	22.1%	annually	3 years	
20. $780.10	$5\frac{3}{4}$ %	annually	4 years	

Discounts and Markups

Complete the last two columns of the chart using the discount rate or markup rate.
Round to the nearest cent.

	Cost/Price	Discount Rate	Markup Rate	Discount or Markup	Sale Price or Selling Price
1.	$35.00	25%			
2.	$42.00		18%		
3.	$68.00		20%		
4.	$24.99	70%			
5.	$50.00		65%		
6.	$20.00	35%			
7.	$17.50	5%			
8.	$110.90		33%		
9.	$240.50	60%			
10.	$89.75		28%		
11.	$64.25	40%			
12.	$19.99		88%		
13.	$595.00		8%		
14.	$616.80	12%			
15.	$200.00		15%		
16.	$450.50	55%			
17.	$38.90	64%			
18.	$14.98		70%		
19.	$5.65		95%		
20.	$717.20		18%		

Commission

Complete the charts. Round to the nearest cent.

	Rate of Commission	Total Sales	Commission		Rate of Commission	Total Sales	Commission
1.	14%	$950.00		21.	$4\frac{7}{8}$ %	$412.13	
2.	22%	$412.75		22.	18%	$5,678.20	
3.	11%	$1,020.80		23.	5.6%	$718.65	
4.	25%	$428.66		24.	28%	$95.25	
5.	15%	$505.15		25.	$12\frac{1}{2}$ %	$648.29	
6.	9%	$3,496.98		26.	33.3%	$300.50	
7.	$10\frac{1}{2}$ %	$54.75		27.	8.2%	$982.17	
8.	30%	$104.73		28.	16%	$1,546.70	
9.	$13\frac{1}{4}$ %	$64.00		29.	$15\frac{1}{4}$ %	$3,009.75	
10.	16%	$89.11		30.	18.5%	$818.40	
11.	35%	$715.25		31.	14%	$335.25	
12.	44%	$300.50		32.	9.6%	$1,124.55	
13.	$8\frac{3}{8}$ %	$2,450.75		33.	12%	$39,428.00	
14.	$11\frac{1}{4}$ %	$918.75		34.	28%	$518.95	
15.	13%	$600.00		35.	$17\frac{3}{4}$ %	$499.99	
16.	24%	$818.95		36.	31%	$4,000.00	
17.	28%	$42.82		37.	42%	$780.99	
18.	$7\frac{1}{2}$ %	$348.60		38.	14.2%	$395.00	
19.	6%	$659.34		39.	14%	$488.62	
20.	5%	$205.12		40.	$8\frac{3}{4}$ %	$199.00	

Down Payment

Complete the charts. Round to the nearest cent.

	Purchase Price	Down Payment Percentage	Down Payment			Purchase Price	Down Payment Percentage	Down Payment
1.	$5,000.00	15%		21.	$750.00	8%	
2.	$1,125.00	20%		22.	$990.95	12%	
3.	$890.50	19%		23.	$4,508.85	20%	
4.	$7,500.00	14%		24.	$1,427.99	15%	
5.	$9,000.00	10%		25.	$843.75	$14\frac{1}{2}$ %	
6.	$1,546.88	5%		26.	$6,000.00	12%	
7.	$2,999.99	25%		27.	$7,300.00	20%	
8.	$8,500.00	50%		28.	$640.25	25%	
9.	$7,400.00	40%		29.	$900.00	40%	
10.	$658.75	35%		30.	$415.50	$11\frac{3}{8}$ %	
11.	$400.00	20%		31.	$21,750.00	22%	
12.	$925.50	15%		32.	$8,173.25	35%	
13.	$717.25	18%		33.	$767.20	$16\frac{3}{4}$ %	
14.	$629.84	15%		34.	$2,480.25	45%	
15.	$3,985.15	10%		35.	$960.00	14.2%	
16.	$200.00	$12\frac{1}{2}$ %		36.	$817.20	20.8%	
17.	$718.99	14%		37.	$415.10	16%	
18.	$515.20	25%		38.	$9,000.00	$9\frac{1}{2}$ %	
19.	$7,600.00	40%		39.	$11,400.00	7%	
20.	$95,000.00	30%		40.	$880.15	6.8%	

Dividing Fractions

Divide. Do your work in the space at the bottom of the page.

1. $\dfrac{8}{9} \div \dfrac{14}{15}$ 2. $\dfrac{3}{10} \div \dfrac{12}{25}$ 3. $\dfrac{7}{50} \div \dfrac{21}{35}$ 4. $\dfrac{11}{18} \div \dfrac{5}{12}$

5. $\dfrac{4}{7} \div \dfrac{8}{9}$ 6. $\dfrac{9}{11} \div \dfrac{2}{3}$ 7. $\dfrac{13}{14} \div \dfrac{1}{2}$ 8. $\dfrac{11}{12} \div \dfrac{5}{8}$

9. $\dfrac{14}{15} \div \dfrac{3}{20}$ 10. $\dfrac{3}{4} \div \dfrac{13}{16}$ 11. $\dfrac{18}{33} \div \dfrac{15}{22}$ 12. $\dfrac{5}{6} \div \dfrac{2}{7}$

13. $\dfrac{3}{10} \div \dfrac{4}{11}$ 14. $\dfrac{5}{12} \div \dfrac{3}{5}$ 15. $\dfrac{17}{40} \div \dfrac{1}{4}$ 16. $\dfrac{4}{5} \div \dfrac{3}{35}$

17. $\dfrac{2}{3} \div \dfrac{5}{8}$ 18. $\dfrac{8}{9} \div \dfrac{11}{12}$ 19. $\dfrac{9}{10} \div \dfrac{7}{9}$ 20. $\dfrac{3}{7} \div \dfrac{12}{13}$

More Dividing Fractions

Divide. Do your work in the space at the bottom of the page.

1. $\dfrac{17}{20} \div \dfrac{4}{5}$ 2. $\dfrac{15}{16} \div \dfrac{5}{14}$ 3. $\dfrac{7}{12} \div \dfrac{2}{11}$ 4. $\dfrac{2}{9} \div \dfrac{4}{11}$

5. $\dfrac{5}{16} \div \dfrac{5}{8}$ 6. $\dfrac{7}{9} \div \dfrac{2}{3}$ 7. $\dfrac{9}{10} \div \dfrac{15}{16}$ 8. $\dfrac{5}{8} \div \dfrac{5}{16}$

9. $\dfrac{14}{15} \div \dfrac{4}{5}$ 10. $\dfrac{22}{35} \div \dfrac{2}{35}$ 11. $\dfrac{2}{7} \div \dfrac{16}{21}$ 12. $\dfrac{21}{40} \div \dfrac{7}{24}$

13. $\dfrac{7}{18} \div \dfrac{5}{16}$ 14. $\dfrac{5}{8} \div \dfrac{4}{5}$ 15. $\dfrac{8}{15} \div \dfrac{12}{25}$ 16. $\dfrac{11}{12} \div \dfrac{1}{6}$

17. $\dfrac{25}{38} \div \dfrac{15}{32}$ 18. $\dfrac{3}{10} \div \dfrac{4}{5}$ 19. $\dfrac{18}{25} \div \dfrac{3}{10}$ 20. $\dfrac{22}{81} \div \dfrac{8}{9}$

Dividing Mixed Numbers

1. $4\frac{2}{7} \div 5\frac{1}{4}$

2. $3\frac{1}{5} \div 1\frac{7}{15}$

3. $1\frac{13}{20} \div 6\frac{7}{8}$

4. $12\frac{1}{2} \div 13\frac{1}{3}$

5. $1\frac{3}{16} \div 1\frac{1}{6}$

6. $1\frac{19}{25} \div 1\frac{1}{10}$

7. $2\frac{14}{15} \div 4\frac{7}{12}$

8. $3\frac{1}{8} \div 16\frac{2}{3}$

9. $4\frac{1}{6} \div 18\frac{1}{3}$

10. $7\frac{1}{3} \div 1\frac{5}{6}$

11. $2\frac{1}{7} \div 8\frac{4}{7}$

12. $11\frac{3}{5} \div 3\frac{1}{15}$

13. $9\frac{1}{6} \div 8\frac{1}{4}$

14. $4\frac{7}{12} \div 2\frac{14}{15}$

15. $2\frac{11}{12} \div 7\frac{7}{8}$

16. $7\frac{7}{10} \div 8\frac{1}{6}$

17. $3\frac{3}{5} \div 1\frac{3}{25}$

18. $5\frac{1}{7} \div 5\frac{1}{3}$

19. $2\frac{11}{20} \div 3\frac{2}{5}$

20. $28\frac{2}{3} \div 2\frac{14}{15}$

Dividing Whole Numbers by Fractions

1. $18 \div 2\frac{1}{4}$

2. $25 \div \frac{5}{13}$

3. $2 \div 1\frac{3}{5}$

4. $44 \div \frac{11}{12}$

5. $21 \div 4\frac{2}{3}$

6. $12 \div 4\frac{4}{5}$

7. $49 \div \frac{21}{45}$

8. $13 \div 8\frac{2}{3}$

9. $18 \div 5\frac{1}{4}$

10. $26 \div 3\frac{1}{4}$

11. $32 \div 2\frac{4}{5}$

12. $27 \div \frac{3}{11}$

13. $75 \div 3\frac{3}{11}$

14. $16 \div 1\frac{1}{5}$

15. $52 \div 2\frac{8}{9}$

16. $24 \div \frac{9}{10}$

17. $64 \div 4\frac{4}{11}$

18. $39 \div 5\frac{1}{4}$

19. $28 \div 1\frac{5}{13}$

20. $64 \div 7\frac{3}{5}$

6.NS.B.2

Divisibility Rules

Check the boxes on the chart that apply.

Divisible by:

Number	2	3	4	5	6	8	9	10
1. 12								
2. 18								
3. 52								
4. 65								
5. 76								
6. 90								
7. 105								
8. 304								
9. 481								
10. 530								
11. 720								
12. 1,342								
13. 2,008								
14. 3,500								
15. 5,896								
16. 1,485								
17. 3,744								
18. 51,840								
19. 15,550								
20. 62,937								
21. 32,768								
22. 59,049								
23. 31,250								
24. 60,480								
25. 98,415								

Dividing by Multiples of 10

Divide. Do your work in the space at the bottom of the page.

1. 72,000,000,000 ÷ 90,000
2. 4,500,000,000 ÷ 500,000
3. 240,000,000 ÷ 40
4. 80,000,000,000 ÷ 20
5. 40,000,000,000 ÷ 2,000
6. 200,000,000 ÷ 40,000
7. 20,000,000,000,000 ÷ 250,000
8. 600,000,000 ÷ 40,000
9. 770,000,000 ÷ 1,100
10. 360,000,000,000 ÷ 1,200,000
11. 3,900,000,000 ÷ 13,000
12. 990,000,000 ÷ 33,000
13. 480,000,000,000 ÷ 40,000
14. 6,000,000,000 ÷ 50,000
15. 7,500,000,000 ÷ 150,000
16. 15,000,000,000 ÷ 750,000
17. 51,000,000 ÷ 30,000
18. 460,000,000 ÷ 200,000
19. 880,000 ÷ 200
20. 4,400,000,000 ÷ 110,000,000
21. 700,000,000 ÷ 1,400,000
22. 75,000,000,000 ÷ 300,000
23. 90,000,000 ÷ 3,000
24. 100,000,000 ÷ 2,500

6.NS.B.2

Dividing Whole Numbers

Divide. Do your work in the space at the bottom of the page.

1. $3,142 \overline{\smash{\big)}6,888,972}$

2. $599 \overline{\smash{\big)}5,236,458}$

3. $4,422 \overline{\smash{\big)}32,603,406}$

4. $6,311 \overline{\smash{\big)}70,113,102}$

5. $973 \overline{\smash{\big)}41,737,808}$

6. $4,911 \overline{\smash{\big)}662,308,915}$

7. $1,514 \overline{\smash{\big)}6,998,004}$

8. $872 \overline{\smash{\big)}81,624,503}$

9. $6,012 \overline{\smash{\big)}58,983,732}$

10. $549 \overline{\smash{\big)}9,032,119}$

11. $2,133 \overline{\smash{\big)}10,681,218}$

12. $439 \overline{\smash{\big)}2,651,560}$

13. $607 \overline{\smash{\big)}7,914,063}$

14. $865 \overline{\smash{\big)}85,637,595}$

15. $718 \overline{\smash{\big)}404,459,452}$

16. $2,111 \overline{\smash{\big)}6,337,899}$

17. $2,877 \overline{\smash{\big)}13,098,981}$

18. $7,143 \overline{\smash{\big)}988,453,612}$

19. $988 \overline{\smash{\big)}70,568,888}$

20. $3,020 \overline{\smash{\big)}654,302,198}$

21. $1,998 \overline{\smash{\big)}14,897,088}$

22. $2,981 \overline{\smash{\big)}342,766,752}$

23. $736 \overline{\smash{\big)}90,112,354}$

24. $3,566 \overline{\smash{\big)}723,056,009}$

More Dividing Whole Numbers

1. 5,466 ⟌ 201,300,444

2. 988 ⟌ 47,188,856

3. 925 ⟌ 193,301,875

4. 1,986 ⟌ 1,004,612,337

5. 5,125 ⟌ 46,140,375

6. 3,111 ⟌ 8,990,119,876

7. 1,111 ⟌ 69,267,517

8. 6,432 ⟌ 47,802,624

9. 881 ⟌ 235,008,512

10. 4,566 ⟌ 5,421,733,612

11. 720 ⟌ 31,999,680

12. 4,019 ⟌ 6,616,008,512

13. 789 ⟌ 77,914,539

14. 3,432 ⟌ 278,517,096

15. 2,153 ⟌ 68,392,114

16. 917 ⟌ 142,699,812

17. 2,099 ⟌ 11,414,362

18. 912 ⟌ 88,723,321

19. 3,202 ⟌ 68,670,092

20. 3,192 ⟌ 161,942,928

Checking Multiplication Using Division

Check the multiplication problems using division. Circle the problems that are correct.

1.
$$
\begin{array}{r}
68{,}727 \\
\times\ \ \ \ \ 449 \\
\hline
30{,}858{,}423
\end{array}
$$

2.
$$
\begin{array}{r}
7{,}805 \\
\times\ \ \ \ \ 663 \\
\hline
5{,}074{,}715
\end{array}
$$

3.
$$
\begin{array}{r}
8{,}617 \\
\times\ \ \ \ \ 38 \\
\hline
337{,}446
\end{array}
$$

4.
$$
\begin{array}{r}
631{,}224 \\
\times\ \ \ \ \ 755 \\
\hline
476{,}574{,}120
\end{array}
$$

5.
$$
\begin{array}{r}
5{,}545 \\
\times\ \ \ \ \ 666 \\
\hline
3{,}692{,}970
\end{array}
$$

6.
$$
\begin{array}{r}
512{,}344 \\
\times\ \ \ \ \ 612 \\
\hline
313{,}554{,}528
\end{array}
$$

7.
$$
\begin{array}{r}
341{,}187 \\
\times\ \ \ \ \ 543 \\
\hline
185{,}605{,}728
\end{array}
$$

8.
$$
\begin{array}{r}
3{,}289 \\
\times\ \ \ \ \ 5{,}877 \\
\hline
19{,}229{,}453
\end{array}
$$

9.
$$
\begin{array}{r}
68{,}723 \\
\times\ \ \ \ \ 3{,}402 \\
\hline
233{,}795{,}646
\end{array}
$$

10.
$$
\begin{array}{r}
18{,}918 \\
\times\ \ \ \ \ 7{,}766 \\
\hline
146{,}917{,}188
\end{array}
$$

11.
$$
\begin{array}{r}
4{,}566 \\
\times\ \ \ \ \ 415 \\
\hline
1{,}794{,}890
\end{array}
$$

12.
$$
\begin{array}{r}
522{,}544 \\
\times\ \ \ \ \ 121 \\
\hline
63{,}227{,}824
\end{array}
$$

13.
$$
\begin{array}{r}
67{,}999 \\
\times\ \ \ \ \ 2{,}140 \\
\hline
145{,}517{,}860
\end{array}
$$

14.
$$
\begin{array}{r}
73{,}464 \\
\times\ \ \ \ \ 865 \\
\hline
64{,}546{,}360
\end{array}
$$

15.
$$
\begin{array}{r}
300{,}019 \\
\times\ \ \ \ \ 185 \\
\hline
55{,}503{,}515
\end{array}
$$

16.
$$
\begin{array}{r}
8{,}114 \\
\times\ \ \ \ \ 9{,}753 \\
\hline
79{,}135{,}842
\end{array}
$$

17.
$$
\begin{array}{r}
67{,}233 \\
\times\ \ \ \ \ 4{,}995 \\
\hline
335{,}828{,}835
\end{array}
$$

18.
$$
\begin{array}{r}
91{,}202 \\
\times\ \ \ \ \ 2{,}333 \\
\hline
212{,}774{,}266
\end{array}
$$

19.
$$
\begin{array}{r}
988{,}662 \\
\times\ \ \ \ \ 3{,}456 \\
\hline
3{,}416{,}815{,}872
\end{array}
$$

20.
$$
\begin{array}{r}
41{,}661 \\
\times\ \ \ \ \ 5{,}702 \\
\hline
237{,}551{,}022
\end{array}
$$

Checking Division Using Multiplication

Check the division problems using multiplication. Circle the problems that are correct.
Remember to add the remainders.

1. 663) 59,825,142 — 90,243

2. 720) 63,428,511 — 88,095 R211

3. 568) 87,044,381 — 153,247 R95

4. 3,421) 22,773,597 — 6,657

5. 1,889) 42,877,657 — 22,698 R1,135

6. 665) 21,944,338 — 32,999 R3

7. 889) 53,240,432 — 59,888

8. 5,021) 64,080,587 — 12,762 R2,585

9. 917) 30,157,379 — 32,877

10. 494) 80,009,006 — 161,961 R272

11. 1,021) 3,845,086 — 3,766

12. 3,456) 28,315,008 — 8,093

13. 863) 24,778,599 — 28,712 R143

14. 921) 77,122,008 — 83,737 R221

15. 754) 68,736,902 — 91,163

16. 2,718) 99,219,877 — 36,504 R2,005

17. 5,426) 35,909,268 — 6,718

18. 2,348) 59,800,686 — 25,468 R1,922

Place Value

In the number 3,024,598,136,670, what is the
place value for:

1= _____

2= _____

3= _____

   and _____

4= _____

5= _____

6= _____   and _____

7= _____

8= _____

9= _____

0= _____   and _____

The number 9,438,722,017,654 has:

_____ trillions

_____ ten-millions   _____ ten-billions   _____ tens

_____ ones   _____ thousands   _____ millions

_____ hundred-thousands   _____ hundreds   _____ hundred-billions

_____ billions   _____ hundred-millions   _____ ten-thousand

What number has:

0 billions	6 trillions	2 ten-billions
3 tens	4 hundred-thousands	7 millions
9 thousands	5 ten-millions	8 hundreds
8 hundred-millions	4 ones	3 ten-thousands
	1 hundred-billion	

The number is ___, ___ ___ ___, ___ ___ ___, ___ ___ ___, ___ ___ ___

In the numbers 5,603,447,628,515 and 6,613,436,528,017, which place values have the same
numbers in each?

_____   _____

_____   _____

_____   _____

Pick the numeral of the specified place values, and fill in the appropriate blanks below.

hundred-millions	6,023,478,929,555	thousands	5,866,111,327,420
tens	8,100,322,568,115	millions	7,897,144,628,404
hundred-billions	4,602,335,777,999	ones	9,991,087,655,222
trillions	3,436,791,199,345	hundred-thousands	8,000,000,432,006
ten-millions	1,998,002,304,511	hundreds	4,444,374,623,632
billions	5,312,344,576,104	ten-thousands	2,577,696,827,328
ten-billions	9,102,004,566,721		

___, ___ ___ ___, ___ ___ ___, ___ ___ ___, ___ ___ ___

Reading and Writing Decimals

Write the word form of each decimal.

1. 2.600836 = _____
   _____
   _____

2. 4,669.3455 = _____
   _____

3. 0.000001 = _____
   _____

4. 11.23406 = _____
   _____

5. 807.005 = _____
   _____

6. 0.94103 = _____
   _____

7. 682.059 = _____
   _____

8. 5.5555 = _____
   _____

9. 0.003603 = _____
   _____

10. 9,167.22 = _____
    _____

Write the correct numeral for each number.

11. one hundred eighteen and four hundred ninety-seven hundred-thousandths _____

12. thirty-nine thousand, seventy-four and eighty-seven hundredths _____

13. five thousand eleven and nine thousandths _____

14. nine hundred and six hundred thirty-two thousandths _____

15. eight hundred ninety thousand, one millionths _____

16. forty thousand, six hundred two millionths _____

17. fourteen and sixty-eight thousand, nine hundred twenty-one millionths _____

18. fifty and one thousand ninety-three ten-thousandths _____

19. three hundred-thousandths _____

20. thirty-three thousand, four hundred thirteen hundred-thousandths _____

21. nine thousand, fifty-six and forty-eight ten-thousandths _____

22. nine and one hundred fifty-two ten-thousandths _____

23. five hundred forty-two and six hundred eighty-eight thousandths _____

24. seventeen and nineteen thousandths _____

25. one and eight hundred seventy-one thousand, eighteen millionths _____

Decimal Place Value

On the following number:

4 , 1 3 9 . 6 5 0 1 2 2

1. Circle the number in the tenth's place.
2. Put a triangle around the number in ten-thousandth's place.
3. Draw lips around the number in the millionth's place.
4. Put a star on the number in the one's place.
5. Make an X over the number that is in the hundred-thousandth's place.
6. Draw a heart around the number in the ten's place.
7. Give the number in the hundred's place a smiley face.
8. Make a sun out of the number in the thousand's place.
9. Draw flower petals around the number in the thousandth's place.
10. Put a check mark over the number in the hundredth's place.

The number 9,113.040872 has:

11. ____ hundredths

12. ____ ten-thousandths   15. _____ ones   18. _____ tenths

13. ____ hundred   16. _____ millionths   19. _____ ten

14. ____ hundred-thousandths 17. _____ thousands   20. _____ thousandths

(21-30) What number has:

4 ten-thousandths   0 ones   9 hundredths

8 thousands   6 tens   2 hundred-thousandths

9 millionths   1 hundred   5 thousandths

                                        3 tenths

The number is ___, ___ ___ ___. ___ ___ ___ ___ ___ ___

Use the numeral that holds the specified place value in each decimal below to create a number where it holds the same place value.

hundred-thousandths	4,062.983216	ones	624.567901
tens	918.42658	thousandths	18.722345
thousands	23,498.00542	tenths	127.156686
hundredths	13,568.006781	millionths	0.643112
ten-thousandths	2.34456	hundreds	91,875.2201

The number is ___, ___ ___ ___. ___ ___ ___ ___ ___ ___

Comparing Decimals

Compare. Write <, > or =.

1. 39.00162 _____ 39.00126
2. 1,427.896205 _____ 1,247.896205
3. 0.057727 _____ 0.057772
4. 463.08888 _____ 463.088888
5. 0.00014 _____ 0.01014
7. 8,223.67313 _____ 8,223.76313
9. 0.0099761 _____ 0.099761
11. 29.043652 _____ 29.43652
13. 111.0101101 _____ 111.0101101
15. 5,145.441545 _____ 5,145.44541516
17. 6.431032 _____ 6.413032
19. 732.145889 _____ 732.14589

6. 6.78799 _____ 6.78979
8. 17.045893 _____ 17.045893
10. 382.18764 _____ 382.18674
12. 991.036497 _____ 919.036437
14. 0.0221221 _____ 0.0212212
16. 26,188.7658 _____ 26,818.7654
18. 92.990913 _____ 92.909913
20. 0.663201 _____ 0.6632

Put in increasing order.

21. 0.4689, 0.4698, 0.4869 _____
22. 3.45572, 3.45725, 3.45427 _____
23. 213.95002, 213.90552, 213.59992 _____
24. 0.0066763, 0.0066766, 0.007606 _____
25. 11.121121, 11.121211, 11.112222 _____
26. 8.899881, 8.89988, 8.89998 _____
27. 0.667321, 0.662731, 0.663726, 0.673216, 0.667231 _____
28. 0.0074432, 0.0074322, 0.00744232, 0.00744322, 0.0074342 _____
29. 3,401,782.0036; 3,401,728.0036; 3,401,782.00361 _____
30. 9.882086, 9.88286, 9.8826802, 9.2888888 _____

Put in decreasing order.

31. 89.07653, 89.70065, 89.70007 _____
32. 6.00498, 6.04009, 6.048809 _____
33. 0.112313, 0.1123201, 0.1123103 _____
34. 0.554545, 0.554554, 0.554455 _____
35. 0.631124, 0.63098, 0.631139 _____
36. 0.98062, 0.981, 0.980609 _____
37. 862.044302, 862.043402, 862.0443201, 862.0044989 _____
38. 0.004609, 0.004069, 0.004906, 0.04609, 0.0469 _____
39. 10,881.188118; 10,880.188818; 10,818.188118; 10,880.188188 _____
40. 47.707669, 47.707696, 47.770669, 47.77009 _____

Name_____

6.NS.B.3

Adding Decimals

Add.

1. 612.05466
+ 80.98802

2. 132.0638
+ 99.542

3. 1.39876
26.8238
+ 0.666895

4. 236.7842
8.99999
+ 5,689.803

5. 626.887762
.000903
+ 44.44444

6. 89.0006
4.55
+ 0.368

7. 638.421
9.99808
+ 76.0039

8. 6.890033
2.689763
+ 5.589922

9. 86.982683
0.4036
+ 9.90832

10. 46.325
29.888
+ 98.634

11. 638.26
39.84
+ 119.08

12. 162.894
334.55
+ 712.203

13. 9.0468
0.3201
+ 14.5066

14. 63.8907
0.0655
+ 111.347

15. 2.368
1.114
+ 3.226

16. 891.05637
673.998
+ 1,487.009834

17. 33.8912
0.7436
+ 0.044

18. 0.368811
0.045332
+ 0.585857

19. 47.0883
681.119956
+ 700.8876

20. 1,211.426789
5,781.330042
+ 6,666.000455

21. 565.889
404.011
+ 300.979

22. 0.663988
0.700707
+ 0.301122

23. 4.327
0.689
+ 15.901

24. 0.1111
0.2222
0.3333
+ 0.3334

25. 28.43765
0.002
1.0934
+ 3.0

26. 18
0.00976
5.703
+ 1,426.0008

27. 162.009
38.468
89.001
+ 3,277.698

28. 0.003486
0.915541
0.2389
+ 0.00687

50

© Carson-Dellosa • CD-704389

Subtracting Decimals

Subtract.

1. 92.6873
  − 89.0099

2. 0.90082
  − 0.89726

3. 588.1188
  − 29.5979

4. 36.04
  − 0.9765

5. 3,427.67
  − 563.008

6. 0.99365
  − 0.112

7. 6.537
  − 0.9688

8. 52.5
  − 50.75

9. 11.111111
  − 10.101010

10. 3.98902
  − 1.19399

11. 3.066666
  − 2.578899

12. 14.73402
  − 8.65211

13. 7.00006
  − 5.12349

14. 0.43683
  − 0.09817

15. 2,657.8532
  1,748.9104

16. 666.78918
  − 423.88009

17. 891.04
  − 763.57

18. 3.467
  − 2.00892

19. 918.5
  − 2.665

20. 4,881.633
  − 9.808

21. 99.123
  − 18.0566

22. 0.773286
  − 0.004499

23. 4.877681
  − 0.119991

24. 33.45688
  − 18.92111

25. 0.116102
  − 0.059333

26. 87.7654
  − 69.88888

27. 386.44444
  − 189.98765

28. 1,100.88
  − 26.999

29. 8.0076
  − 1.116542

30. 0.201
  − 0.11998

31. 0.661132
  − 0.450987

32. 63.00913
  − 58.11119

33. 3.87004
  − 1.98112

34. 19.6389
  − 0.9999

35. 1.6893
  − 0.9109

36. 419.00863
  − 98.74296

Adding and Subtracting Decimals

Solve.

1. 6,183.62
    − 5,812.897

2. 0.883621
    + 9.116379

3. 36.81104
    + 13.18896

4. 5.06073
    + 93.684

5. 818
    − 807.66321

6. 6.891236
    − 5.723445

7. 214.68893
    + 98.51773

8. 26.8
    − 0.98712

9. 0.86102
    − 0.39445

10. 118
    − 106.0043

11. 53.0004
    − 19.5567

12. 82.065791
    + 29.554

13. 1,042.9896
    + 9,751.003654

14. 400.68033
    + 563.77141

15. 9,568 − 8,711.5

16. 4.68623
    + 89.632

17. 666,387.491632
    + 301,444.516339

18. 874.00689
    + 719.53654

19. 69.7864
    + 11.8913

20. 68.4388
    − 29.5595

21. 95 − 1.000041

22. 4,689
    + 93.684

23. 568.007
    − 491.63321

24. 0.798633
    − 0.509752

Multiplying Decimals by Powers of 10

Fill in the graph.

Number	x	1,000	0.01	100	10,000	0.001	10	0.1
1. 27.9								
2. 0.0618								
3. 300.46								
4. 0.55								
5. 23.175								
6. 599.86								
7. 1.246								
8. 0.008								
9. 4,682.7								
10. 7.651								
11. 0.00049								
12. 86.3								
13. 9.72								
14. 0.8657								
15. 15,119.6								
16. 6.003								
17. 577.78								
18. 1,064.89								
19. 0.13246								
20. 3.992								
21. 12,981.4								
22. 0.0053								
23. 74.09								
24. 6.0003								
25. 9,086.5								

6.NS.B.3

Multiplying Decimals

1. 84.63
    x 7.64

2. 19.75
    x 4.3

3. 0.876
    x 0.54

4. 5.33
    x 46.7

5. 0.3443
    x 5.7

6. 99.6
    x 0.42

7. 713.26
    x 4.8

8. 88.3
    x 0.462

9. 2.005
    x 0.97

10. 631.8
    x 3.4

11. 66.5
    x 23.3

12. 84.11
    x 0.76

13. 0.688
    x 0.499

14. 0.5987
    x 0.83

15. 0.4458
    x 0.81

16. 28.9
    x 0.103

17. 16.05
    x 35.6

18. 67.031
    x 0.52

19. 4.8912
    x 3.3

20. 65.7
    x 0.104

More Multiplying Decimals

1.
$$\begin{array}{r} 0.008 \\ \times\ 0.073 \\ \hline \end{array}$$

2.
$$\begin{array}{r} 0.123 \\ \times\ 0.42 \\ \hline \end{array}$$

3.
$$\begin{array}{r} 0.099 \\ \times\ 0.088 \\ \hline \end{array}$$

4.
$$\begin{array}{r} 0.30654 \\ \times\ 0.25 \\ \hline \end{array}$$

5.
$$\begin{array}{r} 0.301 \\ \times\ 0.104 \\ \hline \end{array}$$

6.
$$\begin{array}{r} 0.0399 \\ \times\ 0.16 \\ \hline \end{array}$$

7.
$$\begin{array}{r} 0.2107 \\ \times\ 0.34 \\ \hline \end{array}$$

8.
$$\begin{array}{r} 0.003 \\ \times\ 0.005 \\ \hline \end{array}$$

9.
$$\begin{array}{r} 0.244 \\ \times\ 0.15 \\ \hline \end{array}$$

10.
$$\begin{array}{r} 0.6732 \\ \times\ 0.04 \\ \hline \end{array}$$

11.
$$\begin{array}{r} 0.213 \\ \times\ 0.229 \\ \hline \end{array}$$

12.
$$\begin{array}{r} 0.078 \\ \times\ 0.063 \\ \hline \end{array}$$

13.
$$\begin{array}{r} 0.21312 \\ \times\ 0.44 \\ \hline \end{array}$$

14.
$$\begin{array}{r} 0.526 \\ \times\ 0.12 \\ \hline \end{array}$$

15.
$$\begin{array}{r} 0.01098 \\ \times\ 0.79 \\ \hline \end{array}$$

16.
$$\begin{array}{r} 0.3476 \\ \times\ 0.025 \\ \hline \end{array}$$

17.
$$\begin{array}{r} 0.0566 \\ \times\ 0.07 \\ \hline \end{array}$$

18.
$$\begin{array}{r} 0.188 \\ \times\ 0.188 \\ \hline \end{array}$$

19.
$$\begin{array}{r} 0.903 \\ \times\ 0.106 \\ \hline \end{array}$$

20.
$$\begin{array}{r} 0.00078 \\ \times\ 0.65 \\ \hline \end{array}$$

21.
$$\begin{array}{r} 0.811 \\ \times\ 0.062 \\ \hline \end{array}$$

22.
$$\begin{array}{r} 0.212 \\ \times\ 0.323 \\ \hline \end{array}$$

23.
$$\begin{array}{r} 0.512 \\ \times\ 0.123 \\ \hline \end{array}$$

24.
$$\begin{array}{r} 0.4041 \\ \times\ 0.217 \\ \hline \end{array}$$

Dividing Decimals by Powers of 10

Fill in the graph.

Number	÷	0.01	10,000	100	0.0001	0.1	1,000	0.001
1. 63.8								
2. 0.0092								
3. 718.4								
4. 9.663								
5. 500.6								
6. 0.00081								
7. 8.005								
8. 40.067								
9. 7,100.5								
10. 0.06123								
11. 19.63								
12. 44.441								
13. 1,003.6								
14. 6.022								
15. 0.00055								
16. 21,560.3								
17. 0.1399								
18. 20.441								
19. 7,115.8								
20. 6.8897								
21. 17.099								
22. 321.05								
23. 69.4003								
24. 1.0008								
25. 555.11								

Dividing Decimals

1. $12\overline{)0.894}$

2. $34\overline{)2,318.8}$

3. $76\overline{)446.956}$

4. $42\overline{)294.84}$

5. $9\overline{)366.3}$

6. $55\overline{)39,119.3}$

7. $64\overline{)34.688}$

8. $29\overline{)99.789}$

9. $47\overline{)0.01034}$

10. $18\overline{)3.654}$

11. $81\overline{)530.55}$

12. $37\overline{)37.296}$

13. $52\overline{)10.4364}$

14. $26\overline{)256.62}$

15. $71\overline{)4.473}$

16. $90\overline{)636.3}$

17. $15\overline{)520.5}$

18. $44\overline{)4.3956}$

19. $57\overline{)4400.97}$

20. $70\overline{)11.711}$

21. $68\overline{)115.464}$

22. $31\overline{)6888.2}$

23. $86\overline{)410.22}$

24. $14\overline{)42.0658}$

More Dividing Decimals

1. $30.1\overline{)2{,}362.85}$

2. $0.91\overline{)7.28637}$

3. $7.3\overline{)0.00365}$

4. $1.21\overline{)101.6763}$

5. $0.34\overline{)0.306238}$

6. $0.41\overline{)0.008651}$

7. $5.5\overline{)6.963}$

8. $0.85\overline{)8.024}$

9. $4.05\overline{)0.031185}$

10. $0.75\overline{)67.725}$

11. $1.12\overline{)10.08112}$

12. $0.291\overline{)0.012804}$

13. $0.58\overline{)0.32132}$

14. $0.86\overline{)77.83}$

15. $1.09\overline{)2.0492}$

16. $0.038\overline{)0.2109}$

17. $0.17\overline{)1.3583}$

18. $0.081\overline{)0.00405}$

19. $5.1\overline{)0.0357}$

20. $0.209\overline{)15.8004}$

21. $7.5\overline{)0.06975}$

22. $0.011\overline{)0.003784}$

23. $0.89\overline{)5.0196}$

24. $0.9\overline{)27.9198}$

Precision and Greatest Possible Error

Complete the chart.

Measurement	Precision to the Nearest	GPE	Actual Length
1. 82 hm		0.5 hm	82 hm ± 0.5 hm
2.	decigram	0.5 dg	9 dg ± 0.5 dg
3. 247 L	liter		247 L ± 0.5 L
4. 35 cm	centimeter	0.5 cm	
5. 49 kg		0.5 kg	49 kg ± 0.5 kg
6.	milliliter	0.5 mL	112 mL ± 0.5 mL
7. 6 dam	dekameter		6 dam ± 0.5 dam
8. 75 kg	kilogram	0.5 kg	
9. 86 cL		0.5 cL	86 cL ± 0.5 cL
10.	hectometer	0.5 hm	14 hm ± 0.5 hm
11. 647 mm	millimeter		647 mm ± 0.5 mm
12. 51 mg	milligram	0.5 mg	
13. 33 dL		0.5 dL	33 dL ± 0.5 dL
14.	kilometer	0.5 km	240 km ± 0.5 km
15. 467 cg	centigram		467 cg ± 0.5 cg
16. 21.3 cm	0.1 centimeter	0.05 cm	
17. 346.09 L		0.005 L	346.09 L ± 0.005 L
18.	0.1 centimeter	0.05 cm	3.7 cm ± 0.05 cm
19. 29.88 g	0.01 gram		29.88 g ± 0.005 g
20. 9.2 hL	0.1 hectoliter	0.05 hL	
21. 10.17 dam		0.005 dam	10.17 dam ± 0.005 dam
22.	millimeter	0.5 mm	200 mm ± 0.5 mm
23. 918.01 cm	0.01 centimeter		918.01 cm ± 0.005 cm
24. 63.9 L	0.1 liter	0.05 L	
25. 4.003 g		0.0005 g	4.003 g ± 0.0005 g
26.	0.1 kilogram	0.05 kg	26.4 kg ± 0.05 kg
27. 30.7 m	0.1 meter		30.7 m ± 0.05 m
28. 16.42 kg	0.01 kilogram	0.005 kg	
29. 1.111 kL		0.0005 kL	1.111 kL ± 0.0005 kL
30.	0.01 decimeter	0.005 dm	114.14 dm ± 0.005 dm

Factors, Primes, and Composites

Circle the prime numbers and list all the factors of the composite numbers.

1. 29 _____

2. 30 _____

3. 51 _____

4. 55 _____

5. 60 _____

6. 25 _____

7. 75 _____

8. 18 _____

9. 23 _____

10. 16 _____

11. 17 _____

12. 57 _____

13. 100 _____

14. 12 _____

15. 24 _____

16. 28 _____

17. 36 _____

18. 61 _____

19. 49 _____

20. 77 _____

21. 80 _____

22. 73 _____

23. 64 _____

24. 15 _____

25. 65 _____

26. 50 _____

27. 47 _____

28. 14 _____

29. 97 _____

30. 59 _____

Prime Factorization

Find the prime factorization of the following numbers.

1. 320

2. 500

3. 432

4. 104

5. 352

6. 1,539

7. 1,000

8. 1,372

9. 224

10. 792

11. 858

12. 1,020

13. 1,125

14. 8,624

15. 30,030

16. 3,036

17. 900

18. 3,971

19. 3,375

20. 6,732

21. 296

22. 1,435

23. 5,824

24. 10,404

25. 5,929

26. 16,170

27. 18,711

28. 120,050

Greatest Common Factor

Write the greatest common factor of each set of numbers.

1. 140 and 120 _____

2. 84 and 231 _____

3. 315 and 60 _____

4. 40 and 168 _____

5. 270 and 180 _____

6. 168 and 189 _____

Find the greatest common factor for each pair of numbers.

7. 28, 36

8. 12, 16

9. 9, 15

10. 12, 22

11. 10, 15

12. 10, 20

13. 6, 8

14. 6, 10

15. 3, 9

16. 7, 9

17. 4, 8

18. 5, 7

Find the largest common factor of each group of numbers.

19. 8, 18	20. 27, 9
21. 63, 36	22. 49, 42
23. 27, 18, 36	24. 45, 35, 50
25. 7, 5, 9	26. 16, 12, 24
27. 18, 6, 24	28. 50, 20, 30

Least Common Multiple

Write the LCM for each set of numbers.

1. 6 and 9 _____   **2.** 7 and 2 _____

3. 8 and 3 _____   **4.** 5 and 4 _____

5. 2 and 9 _____   **6.** 9 and 3 _____

Write the LCM for each set of numbers.

7. 5 and 9 _____   **8.** 2 and 3 _____

9. 6 and 9 _____   **10.** 3, 5, and 6 _____

11. 3, 6, and 15 _____   **12.** 4, 6, and 7 _____

13. 4, 6, and 10 _____   **14.** 2, 3, and 12 _____

Write the LCM for each set of numbers.

15. 7 and 4 _____   **16.** 3, 6, and 8 _____

17. 4, 6, and 9 _____   **18.** 2, 4, and 7 _____

19. 3, 4, and 6 _____   **20.** 2, 5, and 9 _____

Write the LCM for each set of numbers.

21. 3, 6, and 11 _____

22. 4, 6, and 12 _____

23. 3, 6, and 7 _____

24. 6 and 12 _____

25. 3 and 13 _____

26. 6 and 9 _____

Greatest Common Factor
and Least Common Multiple

Find the GCF and LCM of each pair of numbers.

1. 25, 45   2. 85, 51   3. 18, 21   4. 72, 26

5. 58, 12   6. 15, 65   7. 66, 44   8. 24, 52

9. 42, 40   10. 84, 68   11. 70, 60   12. 64, 28

13. 35, 77   14. 55, 20   15. 60, 75   16. 32, 24

17. 44, 33   18. 24, 46   19. 76, 57   20. 56, 84

21. 16, 20   22. 55, 22   23. 50, 65   24. 77, 66

25. 50, 30   26. 90, 63   27. 80, 28   28. 81, 18

29. 38, 57   30. 48, 36   31. 35, 50   32. 49, 28

Name_____

Positive and Negative Numbers

What is the most likely temperature for:

1.	ice cream	6°C	-2°C	-35°C
2.	ice skating outside	-3°C	35°C	100°C
3.	playing softball	5°C	25°C	75°C
4.	a cup of hot coffee	40°C	80°C	180°C

Write the opposite of each integer.

5. 76   6. -14   7. -6   8. 25

9. -34   10. -37   11. 14   12. 78

13. -57   14. 29   15. -15   16. 5

6.NS.C.6a. 6.NS.C.6c

Integers

Is the number positive or negative?
Write positive or negative.

1. -44   2. 36   3. 51   4. -113

5. -19   6. 26   7. 93   8. -85

9. -12   10. -71   11. 86   12. 225

13. -5   14. -16   15. 29

Write the opposite of each number.

16. 42   17. -7   18. -12   19. -15   20. 21

21. 106   22. -230   23. -81   24. -60   25. 75

26. -111   27. 525   28. -65   29. -33   30. -2

Each symbol represents a number on the number line. Tell which integer is represented by each symbol.

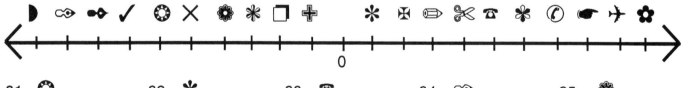

31. ❂   32. ✳   33. ☎   34. ☞   35. ❀

36. ✂   37. ◗   38. ✎   39. ✿   40. ✚

41. ✈   42. ✓   43. ✳   44. ✲   45. ✠

46. ✗   47. ℭ   48. ⚭   49. ☞   50. ❏

Rational Numbers

Write as a rational number in the form $\frac{a}{b}$.

1. 0.6 _____

2. $3\frac{4}{7}$ _____

3. -10 _____

4. -0.82 _____

5. -3.33 _____

6. $-5\frac{5}{6}$ _____

7. 2.12 _____

8. 0.85 _____

9. 27 _____

10. -0.68 _____

11. -9 _____

12. -8.36 _____

13. -4.87 _____

14. $-8\frac{4}{7}$ _____

15. $12\frac{1}{2}$ _____

16. 0.44 _____

17. -0.16 _____

18. 10.3 _____

19. -2.99 _____

20. -0.24 _____

21. 5.25 _____

22. 72 _____

23. $7\frac{3}{10}$ _____

24. 0.28 _____

25. $-6\frac{2}{5}$ _____

Compare. Write >, < or =.

26. $-\frac{3}{5}$ ☐ -0.65

27. $-9\frac{1}{2}$ ☐ $-9\frac{1}{3}$

28. 3.42 ☐ $3\frac{2}{5}$

29. $4.\overline{16}$ ☐ $4\frac{1}{6}$

30. $-0.2\overline{3}$ ☐ $-0.22\overline{3}$

31. $0.4\overline{6}$ ☐ $\frac{7}{16}$

32. $-\frac{2}{9}$ ☐ $-0.\overline{2}$

33. $-\frac{3}{11}$ ☐ $-\frac{2}{5}$

34. -2.2 ☐ -2.22

35. $-\frac{13}{14}$ ☐ $-\frac{13}{16}$

36. $-1\frac{3}{11}$ ☐ $-1.\overline{27}$

37. $-2\frac{5}{8}$ ☐ $-2.\overline{5}$

38. 6.625 ☐ $6\frac{5}{8}$

39. -21 ☐ -21.01

40. $-\frac{3}{7}$ ☐ $-0.\overline{42857}$

Put in increasing order.

41. $12\frac{3}{4}$, -12.74, 12.73, $12\frac{11}{15}$ _____

42. -4.19, -4.201, $-4\frac{2}{9}$, $-4\frac{1}{5}$ _____

43. $\frac{6}{25}$, $\frac{3}{11}$, $\frac{4}{15}$, 0.252 _____

44. $-1\frac{2}{5}$, -1.401, $-1\frac{3}{8}$, 1.389 _____

45. 15.151, $15\frac{3}{16}$, $15\frac{3}{20}$, 15.185 _____

Absolute Value

Evaluate.

1. $|-11|$

2. $|28|$

3. $|33|$

4. $|-110|$

5. $|-50|$

6. $|35|$

7. $|4|$

8. $|-18|$

9. $|-72|$

10. $|-18|$

11. $|-25|$

12. $|-71|$

13. $|-64|$

14. $|44|$

15. $|36|$

16. $|-41|$

17. $|-8|$

18. $|9|$

19. $|214|$

20. $|-510|$

Simplify.

21. $|25| + |-15|$

22. $|-63| - |12|$

23. $|-52| + |-8|$

24. $|-3| \times |-7|$

25. $|24| + |-6|$

26. $|7| \times |-6|$

27. $|-16| + |-9|$

28. $|-7| + |15|$

More Absolute Values

Simplify. Do your work in the space at the bottom of the page.

1. $|43| - |-20|$

2. $|-15| \div |3|$

3. $|5| \times |-4|$

4. $|-20| + |-34|$

5. $|-12| - |-7|$

6. $|-22| \div |-2|$

7. $|-2| \times |20|$

8. $|70| - |-51|$

9. $|-18| \div |-3|$

10. $|-40| - |-17|$

11. $|8| \times |-8|$

12. $|50| \div |-5|$

13. $|36 - 50|$

14. $|-6 \times 5|$

15. $|-7 + -2|$

16. $|-36 \div -6|$

17. $|5 + -10|$

18. $|-11 - -3|$

19. $|40 \div -4|$

20. $|-50 \times -2|$

21. $|-12 + 15|$

22. $|4 \times -7|$

23. $|-81| \times |-16|$

24. $|-301 + 296|$

Properties of Integers

Write which property is used.

1. 8 x (3 + 6) = (8 x 3) + (8 x 6)

2. -55 + 0 = -55

3. -14 x -8 = -8 x -14

4. 8 + (11 + -5) = (8 + 11) + -5

5. 0 + -16 = -16

6. 11 x (-4 + -5) = (11 x -4) + (11 x -5)

7. (-12 x 4) x -3 = -12 x (4 x -3)

8. 66 + -98 = -98 + 66

Use the properties to solve the equations.

9. -36 + x = -36

10. n x -11 = -11 x 9

11. (-11 x -5) + (-11x 6) = n x (-5 + 6)

12. 19 x (n + -4) = (19 x -5) + (19 x -4)

13. 14 + 17 = 17 + n

14. 1 x n = -55

15. (6 + -11) + n = 6 + (-11 + -13)

16. 36 + -15 = n + 36

17. n x -18 = -18

18. 0 + n = -75

19. 8 x (4 + -10) = (8 x n) + (8 x -10)

20. n + (-12 + -13) = (-9 + -12) + -13

Adding Integers With Like Signs

Add.

1. 18 + 11

2. -13 + -11

3. -7 + -12

4. -8 + -6

5. -20 + -4

6. 8 + 13

7. -14 + -6

8. -3 + -9

9. -20 + -2

10. -9 + -7

11. -15 + -6

12. -18 + -3

13. -6 + -7

14. -34 + -5

15. -20 + -13

16. 30 + 30

17. -18 + -4

18. -50 + -5

19. -25 + -16

20. -28 + -9

21. -15 + -21

22. -8 + -80

23. -2 + -19

24. 35 + 62

25. -14 + -9

26. 16 + 24

7.NS.A.1a, 7.NS.A.1d

Adding Integers With Unlike Signs

Add.

1. 11 + -19   2. -4 + 20   3. -18 + 11   4. 26 + -29

5. 14 + -8   6. -48 + 96   7. 81 + -66   8. 28 + -90

9. -42 + 100   10. 88 + -140   11. -16 + 12   12. -90 + 72

13. 14 + -56   14. 20 + -35   15. -28 + 51   16. 17 + -42

17. 28 + -11   18. -53 + 62   19. -40 + 28   20. -93 + 105

21. 42 + -20   22. -80 + 64   23. 59 + -84   24. -4 + 89

25. -18 + 75   26. 71 + -44   27. 92 + -200   28. 22 + -63

29. 16 + -59   30. -94 + 163   31. -303 + 303   32. 422 + -109

Subtracting Integers

Subtract.

1. -70 – 42

2. 18 – -12

3. -38 – 14

4. 49 – -58

5. -33 – -51

6. -11 – 72

7. 28 – -94

8. -54 – -25

9. 75 – 96

10. -81 – 105

11. -35 – 20

12. -150 – 390

13. -39 – -59

14. 50 – 120

15. 14 – -27

16. -98 – -43

17. 77 – -21

18. -60 – 40

19. 8 – -25

20. -3 – 64

21. 21 – -9

22. -8 – 37

23. -120 – -98

24. 105 – -73

25. 50 – -43

26. -62 – -50

Adding and Subtracting Integers

Add and subtract. Do your work in the space at the bottom of the page.

1. -9 + -11 − 4 + -8 − 10

2. 8 − 14 + 95 − -105 + -111 − 63

3. 111 + -128 − -98 − -74 + 110

4. -28 − -43 − 16 − -20 + 89 + -105

5. 89 − -62 − 49 + 68 + 3 − -41

6. 16 − -21 − 28 + -99 − -54 − -17

7. -400 − 32 − -58 + 63 + -94 − 6

8. 48 − 63 + -11 + 25 − -26 + -21

9. 78 − -23 − 49 + 63 + -98 − -19

10. -65 + -94 + 68 − 23 − -89 + -63

11. 36 − -42 + 6 − -28 − 43 − -81 − 6

12. -4 + 8 − 10 − -11 + -13 + 5 − 11 + 12 − -14

More Adding and Subtracting Integers

Add and subtract. Do your work in the space at the bottom of the page.

1. -90 − -27 + 105 − -230 + -64

2. 81 − 104 + 29 − -33 + -56 − 78

3. 42 + 40 + -89 − -64 − 76 + 91

4. -50 − -41 − -65 + 205 − 318 + -5

5. -18 − 29 − -60 + 58 + -70

6. 39 − -82 − 68 + 95 − 53 − -48 + -18

7. -193 − -205 + -68 − 211 − -150

8. 420 − 561 − -502 + 418 + -715 − -42

9. -73 − -68 − 52 + 19 + -105

10. 218 + 195 − 75 + -188 − -163

11. -50 + 77 − -84 − 93 + -60 − -22

12. 409 + -518 − -210 + -68 − -115 + 96

7.NS.A.2a, 7.NS.A.2c

Multiplying Integers

1. $-64{,}515$
   $\times\ \ -980$

2. $70{,}426$
   $\times\ \ \ \ 88$

3. -562
   $\times\ -198$

4. 516
   $\times\ -293$

5. $-3{,}842$
   $\times\ \ \ \ 19$

6. $51{,}826$
   $\times\ \ \ -77$

7. $8{,}265$
   $\times\ -444$

8. $54{,}178$
   $\times\ \ \ 328$

9. -665
   $\times\ -313$

10. -908
    $\times\ 113$

11. $-7{,}268$
    $\times\ \ \ 158$

12. $-60{,}170$
    $\times\ \ \ -425$

13. $9{,}119$
    $\times\ 205$

14. -516
    $\times\ -39$

15. $3{,}009$
    $\times\ -717$

16. $-16{,}412$
    $\times\ \ \ -908$

17. $-28{,}110$
    $\times\ \ \ \ \ 91$

18. $-5{,}648$
    $\times\ \ \ -99$

19. $-2{,}372$
    $\times\ \ \ \ 71$

20. $-30{,}194$
    $\times\ \ \ \ -28$

Dividing Integers

1. 37,668 ÷ -73

2. 56,482 ÷ 62

3. -6,902 ÷ -58

4. 28,576 ÷ -47

5. -6,176 ÷ -32

6. 8,236 ÷ -29

7. 13,536 ÷ -94

8. -56,712 ÷ 51

9. 13,158 ÷ -43

10. -46,260 ÷ -90

11. 33,538 ÷ 82

12. -6,150 ÷ -75

13. -16,188 ÷ -71

14. -9,021 ÷ 93

15. 15,447 ÷ -19

16. -4,624 ÷ 68

17. 6,372 ÷ -59

18. -37,584 ÷ -58

19. 35,388 ÷ 36

20. -29,232 ÷ -48

Multiplying and Dividing Integers

1. -18 x -45 ÷ 54 x -40 ÷ -30

2. 2,400 ÷ -15 x -3 ÷ -60 x 11

3. -85 x -4 ÷ 17 x -22 ÷ 55 x 8

4. -216 ÷ 18 x -13 ÷ 2 x 5 ÷ 39

5. 81 x -15 ÷ 27 x -4 ÷ 60 x -14

6. -270 ÷ -18 x 16 ÷ -15 x -22 ÷ -44

7. 22 x -55 ÷ 10 x -36 ÷ 33 x 4

8. -528 ÷ 24 x -9 ÷ -11 x 50 ÷ -45

9. 28 x -15 ÷ 21 x -35 ÷ 28 x -6

10. -504 ÷ -42 x -33 ÷ 2 x -21 ÷ 77

11. 72 x 11 ÷ -12 x 18 ÷ 4 x 5 ÷ 33

12. 156 ÷ -2 x -5 ÷ 13 x -7 ÷ -35 x -6

13. 33 x -56 ÷ -77 x 25 ÷ -300 x 27 ÷ -18

14. 2,640 ÷ -22 x 9 ÷ 24 x -5 ÷ 15

15. -26 x -20 ÷ 52 x -9 ÷ 5 x -7 ÷ -6

Fractions as Repeating Decimals

Write as repeating decimals.

1. $\dfrac{2}{11}$

2. $\dfrac{5}{6}$

3. $3\dfrac{8}{15}$

4. $2\dfrac{2}{27}$

5. $13\dfrac{11}{18}$

6. $1\dfrac{3}{22}$

7. $\dfrac{5}{33}$

8. $\dfrac{17}{44}$

9. $\dfrac{1}{48}$

10. $6\dfrac{4}{11}$

11. $\dfrac{3}{7}$

12. $\dfrac{23}{30}$

13. $\dfrac{11}{24}$

14. $2\dfrac{2}{15}$

15. $\dfrac{11}{30}$

16. $\dfrac{21}{27}$

17. $\dfrac{2}{37}$

18. $2\dfrac{5}{9}$

19. $108\dfrac{2}{3}$

20. $4\dfrac{1}{36}$

21. $4\dfrac{2}{3}$

22. $14\dfrac{1}{6}$

23. $5\dfrac{17}{33}$

Name_____

More Repeating Decimals

Write as repeating decimals.

1. $1\frac{8}{9}$

2. $\frac{7}{24}$

3. $\frac{7}{45}$

4. $\frac{17}{18}$

5. $\frac{35}{36}$

6. $218\frac{5}{6}$

7. $\frac{5}{66}$

8. $1\frac{1}{99}$

9. $54\frac{8}{11}$

10. $\frac{5}{54}$

11. $\frac{28}{33}$

12. $2\frac{9}{13}$

13. $3\frac{11}{27}$

14. $1\frac{7}{60}$

15. $2\frac{7}{45}$

16. $12\frac{15}{22}$

17. $2\frac{7}{30}$

18. $\frac{5}{18}$

19. $29\frac{2}{9}$

20. $25\frac{1}{3}$

Exponents

Here are three square numbers. Please draw a picture to represent 5 squared or 5^2. Name three square numbers greater than 95.

1.

Write the number represented by each exponent.

2. $5^2 =$ _____

3. $7^3 =$ _____

4. $4^3 =$ _____

5. $8^3 =$ _____

6. $3^4 =$ _____

7. $5^4 =$ _____

8. $3^3 =$ _____

9. $2^5 =$ _____

10. $6^3 =$ _____

Add or subtract.

11. $2^2 + 2^2 =$

12. $3^3 - 2^2 =$

13. $5^2 + 2^3 =$

14. $6^2 + 4^3 =$

15. $4^2 - 2^3 =$

16. $9^3 + 6^2 =$

17. $7^2 - 3^3 =$

18. $8^2 - 3^3 =$

Solve.

19. $9^2 - 8^2 =$

20. $5^3 \div 3 =$

21. $6^2 \div 9 =$

22. $10^2 \div 5 =$

23. $65 - 7^2 =$

24. $7^2 - 3^2 =$

Fractions and Exponents

Solve.

1. $\left(\dfrac{5}{6}\right)^2$

2. $\left(\dfrac{7}{10}\right)^2$

3. $\left(\dfrac{3}{4}\right)^3$

4. $\left(\dfrac{9}{13}\right)^2$

5. $\left(\dfrac{1}{11}\right)^2$

6. $\left(\dfrac{4}{15}\right)^2$

7. $\left(\dfrac{1}{2}\right)^4$

8. $\left(\dfrac{1}{3}\right)^3$

9. $\left(\dfrac{5}{7}\right)^2$

10. $\left(\dfrac{2}{5}\right)^3$

11. $\left(\dfrac{3}{8}\right)^2$

12. $\left(\dfrac{7}{9}\right)^2$

13. $\left(\dfrac{3}{25}\right)^2$

14. $\left(\dfrac{1}{6}\right)^3$

15. $\left(\dfrac{5}{16}\right)^2$

16. $\left(\dfrac{4}{7}\right)^2$

17. $\left(\dfrac{7}{8}\right)^2$

18. $\left(\dfrac{5}{12}\right)^2$

19. $\left(\dfrac{4}{5}\right)^3$

20. $\left(\dfrac{11}{20}\right)^2$

Order of Operations

Work the following problems. Remember the order of operations.

1. $(4 \times 8) \div 2 + 6 =$

2. $16 - 4 + (13 + 7) =$

3. $4 + (9 \div 3) + 3 =$

4. $(100 \div 4) \times 3 =$

5. $9 + 3 - 6 \times 2 \div 4 =$

6. $17 + 7 \times 2 - 11 =$

Remember the order of operations.

7. $(2 \times 3) + 4 \div 2 =$

8. $75 \div 3 \div 5 + 1 =$

9. $33 \times 3 + 1 \div 2 =$

10. $85 \times (7 + 3) \div 5 =$

11. $3 + (21 \div 3) \times 2 =$

12. $4 \times 2 + 5 \times 8 - 2 =$

13. $93 + (24 \div 6) - 4 \div 4 =$

14. $19 - 2 + (34 \div 2) =$

15. $9 \times 6 + (3 \times 2) =$

Solve.

16. $(4 \times 5) - 6 \times 3 + 5 =$

17. $44 \div 2 + 6 \times (2 + 3) =$

18. $4 + (10 \div 2) - 7 =$

19. $26 \div 2 \times (4 \times 3) =$

20. $(8 + 3) \times (17 - 12) =$

More Order of Operations

Solve the following problems. Remember the order of operations.

1. $(6 - 3) + 8 \div 2 =$

2. $5 \times (8 + 4) \div 6 + 3 =$

3. $4 \times 9 - (4 \times 2) =$

4. $5 - 2 + (8 \div 2) \times 3 =$

5. $11 - 5 + 4 \div 2 =$

6. $(8 \times 4) - (4 \times 4) + 4 =$

Solve the following problems. Remember the order of operations.

7. $(6 \times 7) + (4 \times 2) - 25 =$

8. $19 - (3^2) + 32 \div 2 =$

9. $35 + 11 - 6 \div 2 =$

10. $(21 - 6) \times (9 \div 3) + 5 =$

11. $94 - 2 + 3 - (5 \times 5) =$

12. $5 \times 6 \div 3 \div 2 \times 10 =$

Solve the following problems. Remember order of operations.

13. $(3 \times 4) \div (4 - 2) + 4 =$

14. $75 - (5 \times 5) \div 5 =$

15. $(75 - 5) \times 5 \div 5 =$

16. $64 - 16 + 4 \div 10 =$

17. $(6 \times 7) + (3 \times 5) \div 5 =$

18. $6 \times (7 + 3) \times (5 \times 5) =$

Algebraic Expressions

Write the algebraic expression for each word expression.

1. r decreased by 4

_____

2. y divided by 2

_____

3. p and 7

_____

4. d times 3

_____

5. 12 divided by g

_____

6. 5 less than k

_____

Write the algebraic expression for each word expression.

7. 5 more than y

_____

8. 16 minus r

_____

9. 10 divided by p

_____

10. 6 times z

_____

11. m divided by 3

_____

12. 4 more than b

_____

Write the word expression for each algebraic expression.

13. t4 _____

14. 5 + h _____

15. r/5 _____

16. w6 _____

17. q+2 _____

18. 44/f _____

Solve each algebraic expression for the given values.

19. Solve 2g
if g = 2.45 _____   if g = 34 _____   if g = 0.086 _____

20. Solve h – 3.465
if h = 4.02 _____   if h = 56 _____   if h = 6.3 _____

More Algebraic Expressions

Solve for each letter.

1. $3.06 + m + 39.682 = 67.213$ $m = $ _____

2. $.09s = 3.852$ $s = $ _____

3. $k - 1.6429 = 0.4571$ $k = $ _____

4. $325.89/p = 5.1$ $p = $ _____

5. $t + 274,809 = 744,047$ $t = $ _____

6. $8.3a = 18.758$ $a = $ _____

Solve for each letter.   Solve each algebraic expression for the given values.

7. $s = (m + n)(j - k)$

$m + m = j$

$n + n = m$

$n + k = 5$

$k - n = 1$

$s = $ _____ $m = $ _____ $n = $ _____

$j = $ _____ $k = $ _____

8. $6.2 + k \quad = $ _____

$k = 4.97$

9. $\frac{1}{3}s \quad = $ _____

$s = 12$

10. $3\frac{1}{3} - p \quad = $ _____

$p = 1\frac{2}{3}$

Solve for each letter.

11. $a = (b - t) \times a \div t$

$a \div t = 3$

$t + t + t = a$

$t < 3$

$3 < a < 8$

$a = $ _____ $b = $ _____ $t = $ _____

12. $s \div (p - r) - (m \times c) = c$

$c + c = m$

$m + m = p$

$1 + r = p$

$p + c = s$

$3c = 6$

$c = $ _____ $m = $ _____ $p = $ _____ $r = $ _____ $s = $ _____

Solving by Substitution

Evaluate the expressions. Use a=6, b=5 and c=4.

1. $b + 9$

2. $a - c$

3. $a + 8$

4. $\dfrac{12}{c}$

5. $\dfrac{15}{b}$

6. $c - 1$

7. bc

8. $a + -c$

9. $4b - a$

10. $4(b + c)$

11. $8 + c$

12. b^2

13. $c^3 - a^2$

14. $\dfrac{33}{a + b}$

15. $6a + -8c$

16. $ca + b$

17. $\dfrac{26 - a}{c}$

18. $a^2 + b$

19. $\dfrac{2b + 2}{a}$

20. $\dfrac{a^2}{c}$

Evaluate the expressions. Use x=2, y=7 and z=-5.

21. $z - y$

22. $2y + x$

23. xz

24. y^2

25. $4xy - z$

26. $xy + 1$

27. $28 \div y$

28. $6z$

29. $3(x + y)$

30. $2xyz$

31. x^3

32. 3^x

33. 10^x

34. xz^x

35. $\dfrac{y - z}{x}$

36. $\dfrac{16 - x}{y}$

37. $y^x - 6y$

38. $4^x - y$

39. z^x

40. $yz + 40$

Equations

Is the given number a solution of the equation? Write yes or no.

1. $\dfrac{c}{-9} = 8$; 72

2. $8a = -64$; 8

3. $\dfrac{y}{7} = -6$; -42

4. $49 = 7m$; 7

5. $b + 11 = -8$; 19

6. $n - 9 = -6$; 3

7. $-6y = 60$; -10

8. $40 = x - 11$; 29

9. $5 = \dfrac{n}{-21}$; -105

10. $-5 = r + 26$; -33

11. $\dfrac{a}{-13} = -4$; 52

12. $20 = 4n + 12$; 3

13. $3 = c - 22$; 26

14. $12c = 48$; -3

15. $a - 35 = -12$; 23

16. $19 = \dfrac{x}{-5}$; -85

17. $m + 15 = 4$; -11

18. $-75 = 15x$; -3

19. $-99 = -33b$; 3

20. $40 = 5x + 15$; 5

Addition and Subtraction Equations

Solve.

1. $5 + x = 8$

2. $m - 11 = 19$

3. $-6 = -a + 3$

4. $a + 20 = 33$

5. $-4 = 13 + b$

6. $150 + b = 163$

7. $n - 14 = -11$

8. $x + 8 = -5$

9. $21 + c = 30$

10. $18 - x = 3$

11. $-15 = x - 20$

12. $18 = x + 13$

13. $x + -9 = 15$

14. $-15 = c - 7$

15. $44 + x = 56$

16. $19 - c = 11$

17. $n + 14 = -11$

18. $6 = a - 25$

19. $y + 7 = -14$

20. $b + 25 = 4$

21. $b + 9 = -11$

22. $-20 = d - 8$

23. $x - 10 = -8$

24. $-14 + y = -3$

25. $12 - y = 1$

26. $19 = c + 30$

27. $b - 13 = -25$

28. $4 = y - 16$

29. $45 + a = 22$

30. $-20 = a + 11$

31. $13 = 38 + x$

32. $-45 = d - 50$

33. $-14 = a - 39$

34. $8 + y = -30$

35. $11 - y = 5$

36. $15 = -22 + y$

37. $-5 = 16 + x$

38. $a - 63 = 7$

39. $-7 + y = -4$

40. $-16 = c - 22$

Multiplication and Division Equations

Solve.

1. $4x = -20$

2. $\dfrac{n}{6} = 3$

3. $64 = 8y$

4. $11 = \dfrac{a}{-4}$

5. $\dfrac{n}{-14} = 2$

6. $49 = -7x$

7. $-10 = \dfrac{b}{4}$

8. $36 = 4y$

9. $6 = \dfrac{c}{7}$

10. $3a = -45$

11. $\dfrac{x}{-11} = -9$

12. $-5x = 80$

13. $-48 = -12c$

14. $\dfrac{c}{-8} = 9$

15. $7b = -77$

16. $\dfrac{x}{5} = 13$

17. $-8y = 120$

18. $-12 = \dfrac{y}{-6}$

19. $120 = 20n$

20. $\dfrac{n}{-10} = 13$

21. $-8 = \dfrac{y}{11}$

22. $-52 = -13m$

23. $15 = \dfrac{a}{9}$

24. $60 = 6x$

25. $-39 = -3n$

26. $5 = \dfrac{m}{8}$

27. $\dfrac{n}{-12} = -4$

28. $2a = -90$

29. $81 = 9b$

30. $\dfrac{x}{25} = -8$

31. $-10m = 110$

32. $\dfrac{y}{-9} = -9$

33. $12 = \dfrac{x}{7}$

34. $-63 = 21x$

35. $8y = 56$

36. $-4 = \dfrac{m}{20}$

37. $-99 = -11a$

38. $4 = \dfrac{b}{21}$

39. $72 = 9c$

40. $\dfrac{a}{-13} = 2$

Name_____

Equations With Two Operations

Solve. Do your work in the space at the bottom of the page.

1. $3x + 4 = 25$

2. $\frac{x}{4} + 3 = 11$

3. $8n + 21 = -43$

4. $63 = 9a - 27$

5. $8 = 6x - 4$

6. $\frac{p}{-5} + 6 = -5$

7. $-55 = -8n + 9$

8. $5x + 1 = 21$

9. $-20 = -11s + 24$

10. $\frac{r}{8} + 5 = 4$

11. $7x - 11 = 3$

12. $8r - 7 = 17$

13. $20w + 5 = 85$

14. $12c - 16 = 44$

15. $13r - 11 = 28$

16. $\frac{c}{-3} + 16 = -5$

17. $3x - 8 = 28$

18. $\frac{c}{8} + 9 = 15$

19. $-33 = -6r + 9$

20. $7x - 3 = 18$

21. $61 = 16 + 15a$

22. $18 = 22 - \frac{n}{5}$

23. $8 - 2r = -12$

24. $\frac{x}{7} - 5 = 6$

25. $9s + 13 = 85$

26. $22 = 11 + \frac{x}{-4}$

27. $\frac{a}{-5} + 2 = -13$

28. $20b - 93 = 7$

29. $\frac{n}{-12} + 8 = 10$

30. $7a - 28 = 21$

31. $\frac{n}{10} - 3 = 8$

32. $40 - 7x = -16$

33. $-28 = -9x + 17$

34. $16 = \frac{c}{20} + 19$

35. $12r + 33 = 81$

36. $\frac{n}{15} - 13 = -18$

Understanding Area

Find the area., ☐ = 1 square unit. Give your answers in square units.

1.

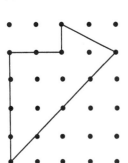

2.

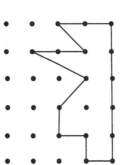

3.

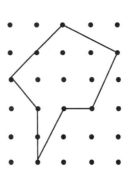

4.

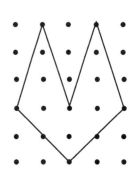

Find the area. ☐ = 1 square unit. Give your answers in square units.

5.

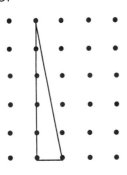

_____

6.

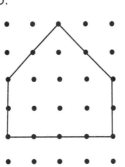

_____

7.

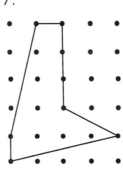

_____

8.

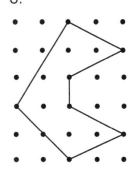

_____

Find the area. ☐ = 1 square unit.

9.

10.

11.

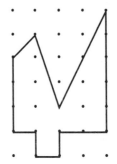

12.

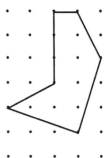

More Understanding Area

Find the area.　☐ = 1 square unit

1.

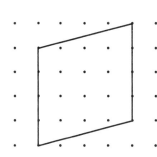

2.

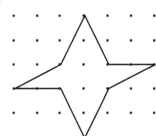

3.

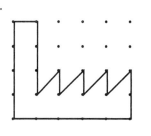

Draw and shade in shapes that have the specified area.

4. 15 sq. units

5.　　10 sq. units

6.　　9 sq. units

Draw and shade in shapes that have the specified area.

7. 13 sq. units

8. 11 sq. units

9. $\frac{1}{2}$ sq. units

Finding Area

Find the area of each figure.

1. 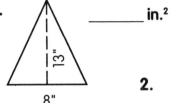 _____ in.²

2.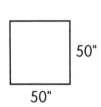

_____ in.²

3.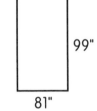

_____ in.²

4.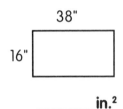

_____ in.²

5.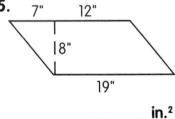

_____ in.²

6.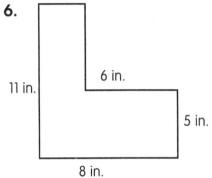

7.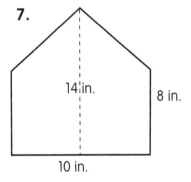

8.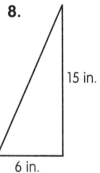

Find the area.

9.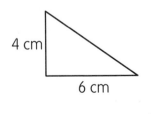

a = _____ cm²

10.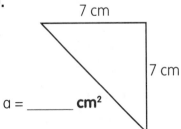

a = _____ cm²

11.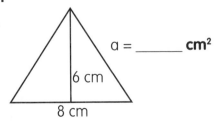

a = _____ cm²

More Practice with Area

Find the area of each figure.

1.

4 cm

10 cm

_____ cm²

2.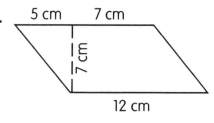

5 cm   7 cm

7 cm

12 cm

_____ cm²

3.

4 cm

6 cm

_____ cm²

4.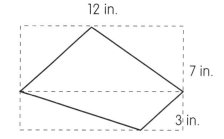

12 in.

7 in.

3 in.

5.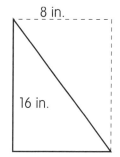

8 in.

16 in.

6.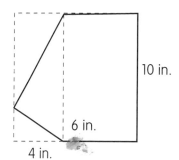

10 in.

6 in.

4 in.

7.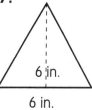

6 in.

6 in.

8.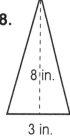

8 in.

3 in.

9.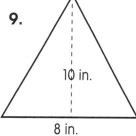

10 in.

8 in.

10.

5 in

2 in.

Surface Area of Rectangular Prisms

Find the surface area of each prism.

1.
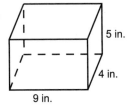
5 in.
4 in.
9 in.

2.
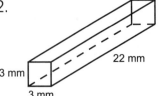
22 mm
3 mm
3 mm

3.
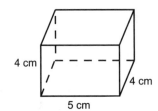
4 cm
4 cm
5 cm

4.
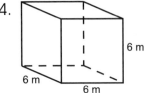
6 m
6 m
6 m

5.
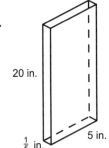
20 in.
$\frac{1}{5}$ in.
5 in.

6.
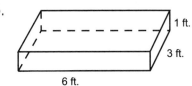
1 ft.
3 ft.
6 ft.

7.
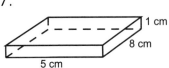
1 cm
8 cm
5 cm

8.
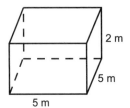
2 m
5 m
5 m

9.
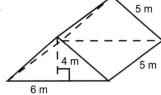
5 m
4 m
5 m
6 m

10.
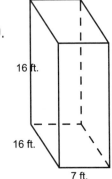
16 ft.
16 ft.
7 ft.

11.
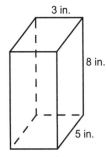
3 in.
8 in.
5 in.

12.
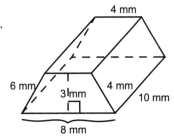
4 mm
6 mm
4 mm
3 mm
10 mm
8 mm

More Surface Area

Find the surface area of each cylinder using the indicated value for pi.

1.

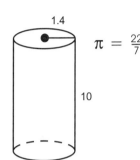

$\pi = \frac{22}{7}$

1.4
10

2.

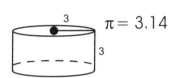

$\pi = 3.14$

3
3

3.

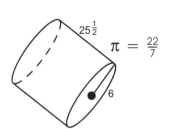

$\pi = \frac{22}{7}$

$25\frac{1}{2}$
6

Find the surface area of each pyramid.

4.

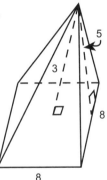

5
3
8
8

5.

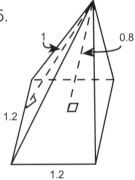

1
0.8
1.2
1.2

6.

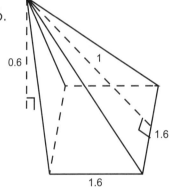

0.6
1
1.6
1.6

Find the surface area of each right circular cone.

7.

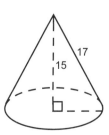

17
15

8.

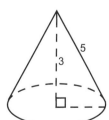

5
3

9.

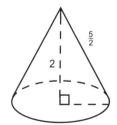

$\frac{5}{2}$
2

Finding Volume

What is the capacity of each shape?

1.
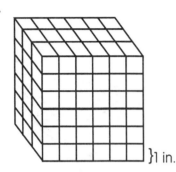
}1 in.

2.
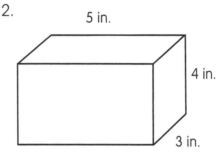
5 in.
4 in.
3 in.

3.
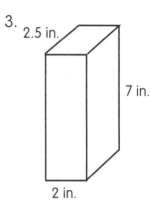
2.5 in.
7 in.
2 in.

Find the volume of each figure.

4. _____

5. _____

6. _____

7. _____ cu. in.

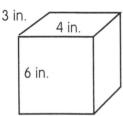
3 in.
4 in.
6 in.

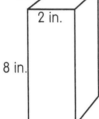
2 in.
2 in.
8 in.

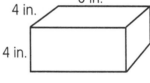
6 in.
4 in.
4 in.

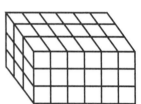

Find the volume of each figure.

8.
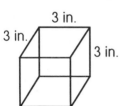
3 in.
3 in.
3 in.
3 in.
_____ in.³

9.
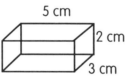
5 cm
2 cm
3 cm
_____ cm³

10.
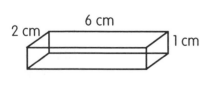
6 cm
2 cm
1 cm
_____ cm³

11.
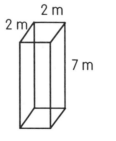
2 m
2 m
7 m
_____ m³

More Practice with Volume

Record the volume of each figure.

1. 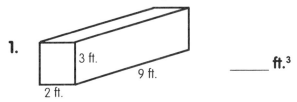 _____ ft.³

2. 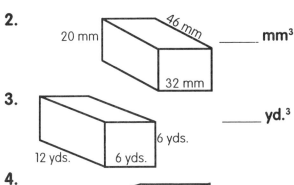 _____ mm³

3. 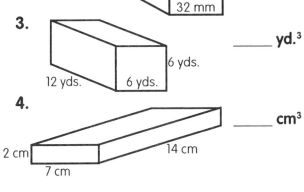 _____ yd.³

4. 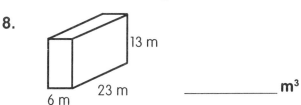 _____ cm³

5. 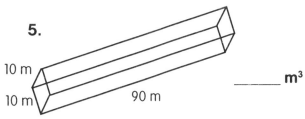 _____ m³

6. 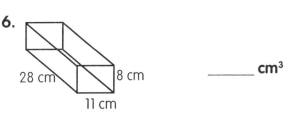 _____ cm³

7. 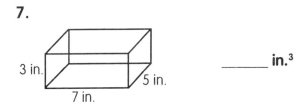 _____ in.³

Record the volume of each figure.

8. 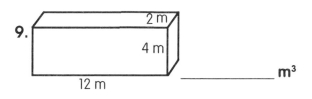 _____ m³

9. 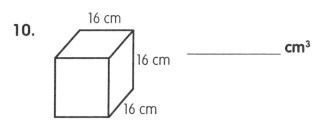 _____ m³

10. _____ cm³

Find the volume.

11. 1 cu. foot = _____ cu. inches

12. 1 cu. yard = _____ cu. feet

13. 2,592 cu. inches = _____ cu. feet

14. 81 cu. feet = _____ cu. yards

15. 1 cu. yard = _____ cu. inches

16. 5,184 cu. inches = _____ cu. feet

Circumference of a Circle

Use the circle at the right to name an example of:

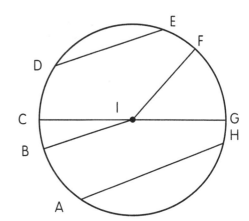

1. chord   5. diameter

2. arc   6. arc

3. center   7. central angle

4. radius

Find the circumference. Use 3.14 for π.

8.

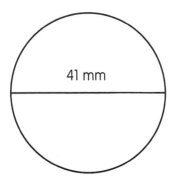

41 mm

9.

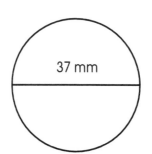

37 mm

10.

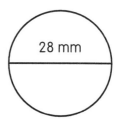

28 mm

Circumference and Area of a Circle

Find the circumference. Use 3.14 for π.

1.

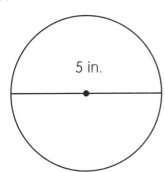

2.

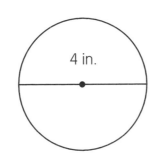

3.

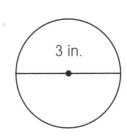

Find the area. Use 3.14 for π.

4.

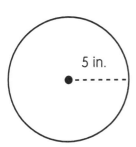

5.

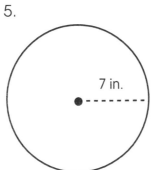

6.

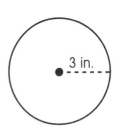

7.

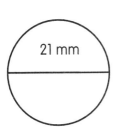

8.

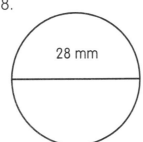

9.

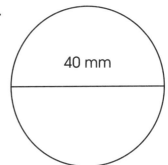

Angles

You may use an angle only once as you name angles that are:

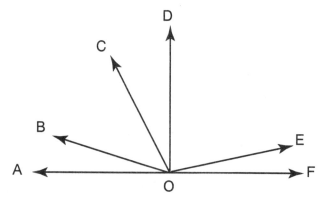

1. acute _____

2. obtuse _____

3. right _____

4. acute _____

5. obtuse _____

Identify each pair of angle as vertical angles, corresponding angles, or supplementary angles.

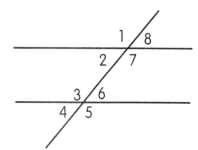

6. ∠7 and ∠2   10. ∠7 and ∠5

7. ∠5 and ∠4   11. ∠3 and ∠1

8. ∠4 and ∠6   12. ∠1 and ∠7

9. ∠7 and ∠1   13. ∠3 and ∠4

Find the missing angle measure in each triangle.

14. 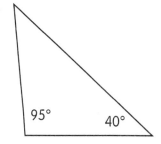   15. 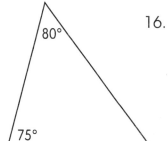   16.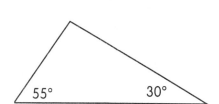

6.RP.A.1, 6.SP.B.5a, 6.SP.B.5b, 7.SP.C.6

Data Distributions

Graph the following data about Yesenia's family vacation.

Number of Miles Traveled

Day	Miles
1	325
2	400
3–5	0
6	220
7	158
8	240
9	110

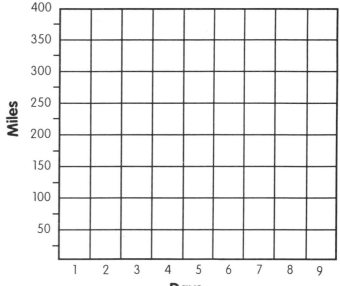

1. How many miles did Yesenia's family travel in all? _____

2. How many miles did they travel on Day 6? _____

3. On which two consecutive days did they travel the farthest?

_____   _____

4. Why do you think they didn't travel at all on days 3–5?

_____

Refer to the table to calculate the relative frequency of each event. Write as a ratio.

Purchases at Movie Night

Grade	Popcorn	Trail Mix	# of Students
Fifth	70	25	95
Sixth	30	50	80
Seventh	45	55	100

What is the relative frequency of . . .

5. a sixth grader purchasing trail mix? _____

6. a fifth grader purchasing popcorn? _____

7. a seventh grader purchasing trail mix? _____

8. a fifth or seventh grader purchasing popcorn? _____

9. a sixth or seventh grader purchasing trail mix? _____

10. a fifth, sixth, or seventh grader purchasing popcorn? _____

Working with Data

Refer to the table to calculate the relative frequency of each event. Write as a ratio.

Favorite Writing Tools

Grade	Pens	Pencils	Markers	Colored Pencils
Fifth	36	52	22	10
Sixth	41	53	11	25
Seventh	63	65	20	12

What is the relative frequency of a . . .

1. seventh grader prefering markers? _____

2. fifth or sixth grader prefering pencils? _____

3. fifth grader prefering pencils or pens? _____

4. sixth or seventh grader prefering pens or markers? _____

5. sixth grader prefering markers or colored pencils? _____

6. fifth, sixth, or seventh grader prefering pens? _____

For three weeks, William recorded his heart rate after running one mile. Use the stem-and-leaf plot to answer the questions.

Heart rate

17	0 2 2 3 4 4 5 7
16	3 5 5 5 5 7 8
15	1 3 8 8 9

7. Which heart rate occurred most often? _____

8. What are the heart rates shown by the first stem and its leaves? _____

9. What is the mode? _____

10. What is the median heart rate? _____

11. Which heart rate occurred more often, 158 or 167? _____

Mean, Mode, Median, Range

Find the mean, mode, median, and range for each set of numbers. Complete the chart.

Set 1.	35	41	68	35	83			
Set 2.	5	12	8	16	12	19		
Set 3.	101	132	100	98	132	124	115	110	132

Set	Mean	Mode	Median	Range
1.				
2.				
3.				

Find the mean, mode, median, and range for each set of numbers. Complete the chart.

Set 4.	5	7	5	9	6	5	5	8	4
Set 5.	26	30	42	55	63	50			
Set 6.	2	93	46	4	16	4	99		

Set	Mean	Mode	Median	Range
4.				
5.				
6.				

Home runs in a season: 38, 42, 36, 17, 38, 21, 24, 38

Find: **7.** the range _____   **8.** the median _____

           9. the mode _____   **10.** the mean _____

Points scored: 19, 17, 18, 19, 16, 6, 10, 12, 9

Find: **11.** the range _____   **12.** the median _____

           13. the mode _____   **14.** the mean _____

6.SP.B.5c, 7.SP.4

More with Mean, Mode, Median, Range

Find the range, median, mean, and mode for each set of numbers.

1. 96, 90, 126, 112, 88, 90, 110, 125, 90

    range = _____   median = _____   mean = _____   mode = _____

2. 585, 501, 399, 313, 424, 476, 501, 568, 355

    range = _____   median = _____   mean = _____   mode = _____

3. 1250, 1315, 1020, 1315, 1442, 1442, 1250, 1442, 1017

    range = _____   median = _____   mean = _____   mode = _____

4. 53, 68, 6, 81, 23, 6, 57, 77, 57, 42, 57, 81, 68

    range = _____   median = _____   mean = _____   mode = _____

5. 94, 97, 98, 99, 94, 91, 95, 98, 90, 95, 98, 92, 96, 96, 92

    range = _____   median = _____   mean = _____   mode = _____

6. 235, 351, 217, 340, 367, 351, 223, 367, 223, 347, 367

    range = _____   median = _____   mean = _____   mode = _____

7. 55,866; 53,866; 57,662; 53,912; 55,866; 52,715; 56,772

    range = _____   median = _____   mean = _____   mode = _____

8. 3, 5, 13, 6, 1, 2, 3, 4, 7, 9, 3, 1, 6, 4, 7, 2, 1, 3, 7, 5, 9, 6, 8

    range = _____   median = _____   mean = _____   mode = _____

9. 100,111; 100,011; 100,001; 100,110; 100,100; 101,000; 100,008; 100,101; 100,110; 100,010; 100,110

    range = _____   median = _____   mean = _____   mode = _____

10. 2033, 2021, 2017, 2035, 2037, 2041, 2029, 2035, 2019, 2017, 2035, 2039, 2019

    range = _____   median = _____   mean = _____   mode = _____

Mean, Mode, Median, Range Challenge

Find the range, median, mean, and mode for each set of numbers.

1. 462,151; 462,511; 462,115; 462,115; 462,511; 462,151; 462,511

    range = _____ median = _____ mean = _____ mode = _____

2. 50,111; 51,105; 50,101; 50,113; 51,110; 51,110; 50,101; 51,115; 51,110

    range = _____ median = _____ mean = _____ mode = _____

3. 1621, 1620, 1611, 1621, 1616, 1605, 1606, 1621, 1613, 1621, 1611

    range = _____ median = _____ mean = _____ mode = _____

4. 310, 311, 315, 307, 310, 301, 305, 301, 313, 315, 311, 310, 303, 313, 310

    range = _____ median = _____ mean = _____ mode = _____

5. 44,040; 40,404; 40,444; 44,004; 44,044; 44,000; 40,400; 44,004; 40,440

    range = _____ median = _____ mean = _____ mode = _____

6. 569, 568, 566, 561, 567, 566, 567, 568, 563, 568, 563

    range = _____ median = _____ mean = _____ mode = _____

7. 21, 27, 11, 36, 38, 40, 44, 35, 23, 19, 19, 39, 21, 25, 33, 44, 43, 33, 19

    range = _____ median = _____ mean = _____ mode = _____

8. 771,711; 771,117; 771,007; 771,077; 771,170; 771,017; 771,171; 771,107; 771,711

    range = _____ median = _____ mean = _____ mode = _____

Probability: Tree Diagrams

Refer to the tree diagrams of possible combinations of doughnut topping choices.

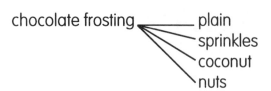
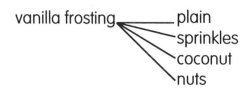

What is the probability of having a doughnut with . . .

1. chocolate frosting and nuts? _____

2. plain frosting? _____

3. frosting with sprinkles? _____

4. chocolate frosting and coconut or nuts? _____

5. vanilla frosting with coconut, nuts, or sprinkles? _____

6. cherry frosting with sprinkles? _____

Draw tree diagrams to illustrate the possible combinations of project choices. Each student must make one choice from each column.

written report	diorama	oral presentation
chapter questions	poster	movement presentation
	game	multimedia presentation
	book	

What is the probability of choosing . . .

7. the written report? _____

8. the game? _____

9. the oral presentation? _____

10. the chapter questions and a movement presentation? _____

11. the diorama and a multimedia presentation? _____

12. the written report, a book, and an oral presentation? _____

Probability

Refer to the table to calculate the probability of each event. Write as a ratio.

Study Time Preferences at Lakeshore School

Grade	Before School	Study Hall	After School	Evening
Fifth	16	82	51	41
Sixth	29	55	54	62
Seventh	31	84	42	53

What is the probability that a . . .

1. sixth grader prefers studying before school? _____

2. sixth or seventh grader prefers studying in study hall? _____

3. fifth grader prefers studying in the evening? _____

4. fifth or seventh grader prefers studying after school?_____

5. seventh grader prefers studying after school or in the evening? _____

6. fifth, sixth, or seventh grader prefers studying before school or in the evening? _____

Use the spinner to determine the probability of each event.

What is the probability of the spinner . . .

7. stopping on an 8? _____

8. stopping on a number less than 14? _____

9. stopping on a number greater than or equal to 6? _____

10. stopping on an even number? _____

11. stopping on an odd number? _____

12. stopping on 6, 8, 10, 14, or 16? _____

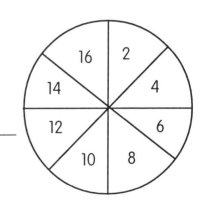

Name_____

7.SP.C.5, 7.SP.C.8a, 7.SP.C.8b

More Probability

You have a number cube with the numbers 1, 2, 3, 4, 5, and 6 on the faces. If you roll it once, what is the probability of getting:

1. the number 3? _____
2. a number other than 4? _____

3. an odd number? _____
4. a number smaller than 6? _____

5. the number 2? _____
6. a number greater than 2? _____

7. an even number? _____
8. a number greater than 1? _____

9. the numbers 4 or 5?_____
10. the numbers 1, 2, or 3? _____

You have a sack with two yellow marbles, three green marbles, one blue marble, and two red marbles. With one draw, what is the probability of getting:

11. a blue marble? _____
12. a red or blue marble? _____

13. a yellow marble? _____
14. a blue or green marble? _____

15. a red marble? _____
16. a red or green marble? _____

17. a green marble? _____
18. a blue or yellow marble? _____

19. any marble but blue? _____
20. any marble but yellow? _____

© Carson-Dellosa • CD-704389

Answer Key

Page 7

1.

2.

↑green

3.

4.
↑yellow

5. $\frac{2}{3}$, 2:3

6. $\frac{9}{6}$, 9:6

7. 7 to 3, 7:3

8. 5 to 9, 5:9

9. 3 to 6, $\frac{3}{6}$

10. 8 to 4, $\frac{8}{4}$

11. $\frac{12}{9}$, 12:9

12. $\frac{1}{7}$, 1:7

13. 8 to 3, $\frac{8}{3}$

14. 12 to 5, 12:5

Page 8

	Ratios		Are ratios equal?	3 ways to write ratios		
1.	$\frac{12}{28}$	$\frac{21}{49}$	✔	$\frac{3}{7}$	3 : 7	0.428571
2.	$\frac{49}{70}$	$\frac{35}{50}$	✓	7/10	7:10	.7
3.	$\frac{6}{24}$	$\frac{5}{20}$	✓	1/4	1:4	.25
4.	$\frac{45}{55}$	$\frac{18}{22}$	✓	9/11	9:11	.8̄1̄
5.	$\frac{2}{12}$	$\frac{9}{54}$	✓	1/6	1:6	.1̄6̄
6.	$\frac{85}{136}$	$\frac{15}{24}$	✓	5/8	5 : 8	.625
7.	$\frac{12}{45}$	$\frac{28}{105}$	✓	4/15	4:15	.2̄6̄
8.	$\frac{18}{64}$	$\frac{22}{98}$				
9.	$\frac{85}{125}$	$\frac{51}{75}$	✓	17/25	17:25	.68
10.	$\frac{42}{63}$	$\frac{30}{45}$	✓	2/3	2 : 3	.6̄
11.	$\frac{39}{59}$	$\frac{52}{120}$				
12.	$\frac{52}{80}$	$\frac{78}{120}$	✓	13/20	13:20	.65
13.	$\frac{36}{45}$	$\frac{28}{35}$	✓	4/5	4:5	.8
14.	$\frac{21}{48}$	$\frac{49}{112}$	✓	7/16	7:16	.4375
15.	$\frac{32}{72}$	$\frac{12}{27}$	✓	4/9	4:9	.4̄

Page 9

1. 55 miles
2. 78 beats
3. 28¢
4. $\frac{3}{4}$ inch
5. 48 words
6. approximately 17¢
7. 26¢
8. 1,900 calories
9. $4.50
10. $1.35

Page 10

1. 14 m per hour
2. $4 per book
3. 500 km per day
4. 22 in. per day
5. $42 per dress
6. 3 dogs per owner
7. 17 m per second
8. 11 km per day
9. 5 in. per hour
10. $300 per ticket
11. 9 cm per second
12. 26 m per hour
13. 21 m per minute
14. 15 toys per baby
15. 2 exams per day
16. 8 m per day
17. $6 per sandwich
18. 14 km per hour
19. 4 travelers per car
20. $125 per coat

Page 11

1. 25 m per hour
2. 4 notebooks per student
3. 12 km per d
4. 18 km per hour
5. 8 candy bars per student
6. $13 per CD
7. 3 in. per second
8. 7 km per hour
9. 22 ft. per minute
10. 6 marbles per child
11. $2 per hotdog
12. 3 in. per second
13. 9 km per hour
14. $7 per pizza
15. 11 m per day
16. 19 km per hour
17. $13,000 per car
18. 7 m per second
19. 6 classes per day
20. $340 per TV

Page 12

1. yes
2. no
3. no
4. yes
5. equivalent
6. equivalent
7. equivalent
8. not equivalent
9. not equivalent
10. equivalent
11. not equivalent
12. equivalent
13. equivalent
14. not equivalent
15. equivalent
16. equivalent
17. equivalent
18. equivalent
19. not equivalent
20. not equivalent
21. not equivalent
22. equivalent
23. equivalent
24. not equivalent

Answer Key

Page 13
1. 168
2. 7
3. 6
4. 7.8
5. 5.4
6. 7.5
7. 56
8. 1.5
9. 9
10. 117
11. 26
12. 96
13. 68
14. $33\frac{1}{3}$%
15. 20
16. 56
17. 216
18. 75
19. 44%
20. 250%
21. 20
22. 600

Page 14
1. 90
2. 17.6
3. 15.6
4. 2
5. 18
6. 63
7. 903
8. 44
9. .64
10. 13.6
11. 15.4
12. 1.2
13. 3.5
14. 4.4
15. 7.4
16. 276
17. 26.25
18. 68.02
19. 18
20. 225

Page 15
1. 40%
2. 312.5%
3. 15%
4. 60.5%
5. 125%
6. 1.2%
7. 18.75%
8. 137.5%
9. 116%
10. 52%
11. 70%
12. 34.375%
13. 55.5%
14. 20%
15. 241.25%
16. 87.5%
17. 77.2%
18. 37.5%
19. 37.5%
20. 20%
21. 250%
22. 6.25%
23. 12%
24. 55%

Page 16
1. 80
2. 80
3. 66
4. 60
5. 96
6. 115
7. 200
8. 120
9. 150
10. 40
11. 280
12. 20
13. 205
14. 1,200
15. 400
16. 75
17. 90
18. 40
19. 160
20. 80

Page 17
1. 400%
2. 56
3. 124.08
4. 96%
5. 37.5%
6. 78.3
7. 24
8. 60%
9. 481.25%
10. 135.6
11. 35
12. 76.8
13. 85%
14. 27.5%
15. 122.2
16. 89.25
17. 62.5%
18. 175.5
19. 14.4
20. 12.5%

Page 18
1. 34
2. 20%
3. 140
4. 30%
5. 28%
6. 392
7. 550
8. 180%
9. 6
10. 80%
11. 99%
12. 68
13. 60
14. 19
15. 36%
16. 92
17. 88%
18. 30%
19. 385
20. 650
21. 35%
22. 95%
23. 20%
24. 40
25. 32%
26. 52

Page 19
1. 40%
2. 48%
3. 85%
4. 34.375%
5. 9.2%
6. 68.75%
7. 60%
8. 4%
9. 43.75%
10. 56.25%
11. 70%
12. 52.5%
13. 62.5%
14. 68%
15. 87.5%
16. 66.6%
17. 82.5%
18. 46%
19. 26.25%
20. 5.5%

Page 20
1. 0.60, 60%, $\frac{3}{5}$
2. 0.38, 38%, $\frac{19}{50}$
3. 0.15, 15%, $\frac{3}{20}$
4. 0.90, 90%, $\frac{9}{10}$
5. 0.59, 59%, $\frac{59}{100}$
6. 0.21, 21%, $\frac{21}{100}$
7. 0.39, 39%, $\frac{39}{100}$
8. 0.02, 2%, $\frac{1}{50}$

Answer Key

9. $0.51, 51\%, \frac{51}{100}$

10. $0.80, 80\%, \frac{4}{5}$

11. $0.12, 12\%, \frac{3}{25}$

12. $0.83, 83\%, \frac{83}{100}$

13. $0.25, 25\%, \frac{1}{4}$

14. $0.06, 6\%, \frac{3}{50}$

15. $0.72, 72\%, \frac{18}{25}$

16. $0.12, 12\%, \frac{3}{25}$

Page 21
1. 0.4
2. 0.09
3. $\frac{75}{100}$ or $\frac{3}{4}$
4. 0.25
5. $\frac{6}{10}$ or $\frac{3}{5}$
6. 0.5
7. $\frac{2}{10}$ or $\frac{1}{5}$
8. $\frac{9}{100}$
9. 0.75
10. $\frac{2}{10}$ or $\frac{1}{5}$
11. 0.3
12. $\frac{83}{100}$
13. $\frac{24}{25}$
14. $1\frac{3}{4}$
15. 50%
16. $\frac{1}{3}$
17. 37.5%
18. 8%
19. $\frac{11}{10}$
20. 40%
21. $\frac{1}{40}$
22. 103%
23. 62.5%
24. $2\frac{3}{4}$

Page 22

	Percent	Fraction	Decimal		Percent	Fraction	Decimal
1.	187.5%	$1\frac{7}{8}$	1.875	21.	19%	$\frac{19}{100}$	.19
2.	18%	$\frac{9}{50}$	0.18	22.	87.5%	$\frac{7}{8}$	.875
3.	2%	$\frac{1}{50}$	.02	23.	180%	$1\frac{4}{5}$	1.8
4.	85%	$\frac{17}{20}$	.85	24.	128%	$1\frac{7}{25}$	1.28
5.	120%	$1\frac{1}{5}$	1.2	25.	5%	$\frac{1}{20}$	.05
6.	135%	$1\frac{7}{20}$	1.35	26.	13.5%	$\frac{27}{200}$	0.135
7.	20.4%	$\frac{51}{250}$	0.204	27.	86.5%	$\frac{173}{200}$	.865
8.	44.5%	$\frac{89}{200}$	.445	28.	290%	$2\frac{9}{10}$	2.9
9.	33%	$\frac{33}{100}$	0.33	29.	352%	$3\frac{13}{25}$	3.52
10.	77%	$\frac{77}{100}$	.77	30.	81.25%	$\frac{13}{16}$	0.8125
11.	40%	$\frac{2}{5}$	.4	31.	72%	$\frac{18}{25}$	.72
12.	42.5%	$\frac{17}{40}$	.425	32.	144%	$1\frac{11}{25}$	1.44
13.	5.5%	$\frac{11}{200}$	.055	33.	61%	$\frac{61}{100}$	.61
14.	195%	$1\frac{19}{20}$	1.95	34.	13.2%	$\frac{33}{250}$	0.132
15.	394%	$3\frac{47}{50}$	3.94	35.	27.6%	$\frac{69}{250}$	.276
16.	264%	$2\frac{16}{25}$	2.64	36.	75.5%	$\frac{151}{200}$	0.755
17.	88%	$\frac{22}{25}$	0.88	37.	45%	$\frac{9}{20}$	0.45
18.	30%	$\frac{3}{10}$	.3	38.	87.5%	$\frac{7}{8}$	.875
19.	12%	$\frac{3}{25}$	.12	39.	27.75%	$\frac{111}{400}$	.2775
20.	71.75%	$\frac{287}{400}$	0.7175	40.	85.8%	$\frac{429}{500}$	0.858

Page 23
1. 44%
2. 56%
3. 8%
4. 84%
5. 21%
6. 67%
7. 88%
8. 57%
9. 53%
10. 73%
11. 88.9%
12. 48.9%
13. 57.1%
14. 41.7%
15. 96.3%
16. 43.3%
17. 69.4%
18. 31.3%
19. 25.3%
20. 94.4%

Page 24
1. 46.67%
2. 27.27%
3. 27.78%
4. 50.63%
5. 46.15%
6. 47.73%
7. 22.22%
8. 80.83%
9. 38.1%
10. 85.71%
11. 26.316%
12. 46.429%
13. 19.231%
14. 54.167%
15. 19.048%
16. 91.667%
17. 28.571%
18. 15.385%
19. 33.333%
20. 90.909%

Answer Key

Page 25

1. 64
2. 64
3. 10
4. 5
5. $1\frac{1}{2}$
6. 4
7. 36 in.
8. 54 in.
9. 24 in.
10. 84 in.
11. 54 in.
12. 234 in.
13. 29 in.
14. 49 in.
15. 19
16. 28
17. $2\frac{1}{4}$
18. 64
19. 14
20. 64

Page 26

	kilometer (km)	hectometer (hm)	dekameter (dam)	meter (m)	decimeter (dm)	centimeter (cm)	millimeter (mm)
1.	.0052	.052	.52	5.2	52	520	5,200
2.	.00367	.0367	.367	3.67	36.7	367	3,670
3.	.000103	.00103	.0103	.103	1.03	10.3	103
4.	.61	6.1	61	610	6,100	61,000	610,000
5.	.008856	.08856	.8856	8.856	88.56	885.6	8,856
6.	.00032	.0032	.032	.32	3.2	32	320
7.	.73	7.3	73	730	7,300	73,000	730,000
8.	.04069	.4069	4.069	40.69	406.9	4,069	40.690
9.	.0000038	.000038	.00038	.0038	.038	.38	3.8
10.	4.4	44	440	4,400	44,000	440,000	4,400,000
11.	.016	.16	1.6	16	160	1,600	16,000
12.	.916	9.16	91.6	916	9,160	91,600	916,000
13.	.0000005	.000005	.00005	.0005	.005	.05	.5
14.	.01007	.1007	1.007	10.07	100.7	1,007	10,070
15.	.142	1.42	14.2	142	1,420	14,200	142,000
16.	.0000082	.000082	.00082	.0082	.082	.82	8.2
17.	.000111	.00111	.0111	.111	1.11	11.11	111.1
18.	.000194	.00194	.0194	.194	1.94	19.4	194
19.	.3	3	30	300	3,000	30,000	300,000
20.	.0000764	.0007641	.007641	.07641	.7641	7.641	76.41
21.	90	900	9,000	90,000	900,000	9,000,000	90,000,000
22.	.00703	.0703	.703	7.03	70.3	703	7,030
23.	.000118	.00118	.0118	.118	1.18	11.8	118
24.	.005	.05	.5	5	50	500	5,000
25.	.0000645	.000645	.00645	.0645	.645	6.45	64.5

Page 27

1. 9
2. 5
3. 15
4. 55
5. 12
6. 16
7. 12
8. 10
9. 100
10. 4
11. 8
12. 9
13. 12
14. 2.7
15. 24
16. 11
17. 470
18. 6
19. 5
20. 12.5
21. 3
22. 42
23. 22.5
24. 15
25. 10.125
26. 6.67
27. 6
28. 14
29. 6
30. 8.75
31. 7.7
32. 4.4

Page 28

1. 80
2. 70
3. 176
4. 16
5. 42
6. 95
7. 5
8. 63
9. 105
10. 30
11. 42
12. 248
13. 9
14. 165
15. 19
16. 176
17. 7
18. 66
19. 18
20. 72

Page 29

1. 19
2. 10
3. 42
4. 63
5. 40
6. 6
7. 40
8. 55
9. 102
10. 36
11. 21
12. 15
13. 125
14. 35
15. 99
16. 48
17. 3
18. 196
19. 9
20. 56

Page 30

	Cost of Item	% Sales Tax	Tax Paid	Total Cost
1.	$4.99	$3\frac{1}{4}$ %	$ .16	$ 5.15
2.	$12.50	5.65%	.71	13.21
3.	$.58	$6\frac{3}{4}$ %	.04	.62
4.	$372.48	12%	44.70	417.18
5.	$111.20	$18\frac{1}{8}$ %	20.16	131.36
6.	$13.84	4.3%	.60	14.44
7.	$25.25	7.11%	1.80	27.05
8.	$30.18	$8\frac{5}{8}$ %	2.60	32.78
9.	$441.89	9.0625%	40.05	481.94
10.	$580.60	14%	81.28	661.88
11.	$14.12	1.35%	.19	14.31
12.	$8.19	6.8%	.56	8.75
13.	$5.45	$5\frac{1}{4}$ %	.29	5.74
14.	$613.20	22%	134.90	748.10
15.	$125.50	$11\frac{3}{8}$ %	14.28	139.78
16.	$220.16	$9\frac{1}{2}$ %	20.92	241.08
17.	$8.12	2.625%	.21	8.33
18.	$9.00	8.9375%	.80	9.80
19.	$16.85	19%	3.20	20.05
20.	$21.22	5.0375%	1.07	22.29

Answer Key

Page 31

	Principal	Interest Rate Per Year	Time	Interest Earned
1.	$625.00	16%	6 months	$ 50.00
2.	$720.50	$7\frac{1}{2}$ %	1 year	54.04
3.	$5,670.80	22%	9 months	935.68
4.	$4,112.20	$11\frac{1}{8}$ %	$4\frac{1}{4}$ years	1944.30
5.	$905.60	14%	$5\frac{1}{2}$ years	697.31
6.	$814.75	$5\frac{3}{4}$ %	4 years	187.39
7.	$1,100.50	15%	3 months	41.27
8.	$870.20	$8\frac{3}{8}$ %	$9\frac{3}{4}$ years	710.57
9.	$415.15	$6\frac{1}{2}$ %	5 months	11.24
10.	$6,540.50	11%	$1\frac{1}{4}$ years	899.32
11.	$11,140.25	5.0375%	8 years	4489.52
12.	$26,500.75	8%	6 months	1060.03
13.	$408.50	2.625%	4 months	3.57
14.	$910.80	21%	3 years	573.80
15.	$12,540.00	$14\frac{5}{8}$ %	7 months	1069.82
16.	$9,750.50	12.0625%	$11\frac{1}{2}$ years	13,525.77
17.	$810.40	$10\frac{1}{2}$ %	6 years	510.55
18.	$4,480.10	4.6875%	3 months	52.50
19.	$33,500.00	33%	$4\frac{1}{4}$ years	46,983.75
20.	$18,549.99	9.6%	5 years	8,964.00

Page 32

1. $511.38
2. $53.58
3. $4,439.35
4. $17.43
5. $12,729.43
6. $432.09
7. $3,800.69
8. $4,830.14
9. $519.10
10. $270.64
11. $325.00
12. $30.21
13. $264.15
14. $914.81
15. $262.07
16. $6,661.21
17. $6,092.00
18. $3,136.44
19. $5,274.68
20. $195.50

Page 33

	Cost/Price	Discount Rate	Markup Rate	Discount or Markup	Sale Price or Selling Price
1.	$35.00	25%		$ 8.75	$ 26.25
2.	$42.00		18%	7.56	49.56
3.	$68.00		20%	13.60	81.60
4.	$24.99	70%		17.49	7.50
5.	$50.00		65%	32.50	82.50
6.	$20.00	35%		7.00	13.00
7.	$17.50	5%		.88	16.62
8.	$110.90		33%	36.60	147.50
9.	$240.50	60%		144.30	96.20
10.	$89.75		28%	25.13	114.88
11.	$64.25	40%		25.70	38.55
12.	$19.99		88%	17.59	37.58
13.	$595.00		8%	47.60	642.60
14.	$616.80	12%		74.02	542.78
15.	$200.00		15%	30.00	230.00
16.	$450.50	55%		247.78	202.72
17.	$38.90	64%		24.90	14.00
18.	$14.98		70%	10.49	25.47
19.	$5.65		95%	5.37	11.02
20.	$717.20		18%	129.10	846.30

Page 34

	Rate of Commission	Total Sales	Commission		Rate of Commission	Total Sales	Commission
1.	14%	$950.00	$ 133	21.	$4\frac{7}{8}$ %	$412.13	20.09
2.	22%	$412.75	90.81	22.	18%	$5,678.20	1022.08
3.	11%	$1,020.80	112.29	23.	5.6%	$718.65	40.24
4.	25%	$428.66	107.17	24.	28%	$95.25	26.67
5.	15%	$505.15	75.77	25.	$12\frac{1}{2}$ %	$648.29	81.04
6.	9%	$3,496.98	314.73	26.	33.3%	$300.50	100.07
7.	$10\frac{1}{2}$ %	$54.75	5.75	27.	8.2%	$982.17	80.54
8.	30%	$104.73	31.42	28.	16%	$1,546.70	247.47
9.	$13\frac{1}{4}$ %	$64.00	8.48	29.	$15\frac{1}{4}$ %	$3,009.75	458.99
10.	16%	$89.11	14.26	30.	18.5%	$818.40	151.40
11.	35%	$715.25	250.34	31.	14%	$335.25	46.94
12.	44%	$300.50	132.22	32.	9.6%	$1,124.55	107.96
13.	$8\frac{3}{8}$ %	$2,450.75	205.25	33.	12%	$39,428.00	4731.36
14.	$11\frac{1}{4}$ %	$918.75	103.36	34.	28%	$518.95	145.31
15.	13%	$600.00	78	35.	$17\frac{3}{4}$ %	$499.99	88.75
16.	24%	$818.95	196.55	36.	31%	$4,000.00	1240
17.	28%	$42.82	11.99	37.	42%	$780.99	328.02
18.	$7\frac{1}{2}$ %	$348.60	26.15	38.	14.2%	$395.00	56.09
19.	6%	$659.34	39.56	39.	14%	$488.62	68.41
20.	5%	$205.12	10.26	40.	$8\frac{3}{4}$ %	$199.00	17.41

Answer Key

Page 35

	Purchase Price	Down Payment Percentage	Down Payment		Purchase Price	Down Payment Percentage	Down Payment
1.	$5,000.00	15%	$750	21.	$750.00	8%	$60
2.	$1,125.00	20%	225	22.	$990.95	12%	118.91
3.	$890.50	19%	169.20	23.	$4,508.85	20%	901.77
4.	$7,500.00	14%	1050	24.	$1,427.99	15%	214.20
5.	$9,000.00	10%	900	25.	$843.75	14½%	122.34
6.	$1,546.88	5%	77.34	26.	$6,000.00	12%	720
7.	$2,999.99	25%	750	27.	$7,300.00	20%	1460
8.	$8,500.00	50%	4250	28.	$640.25	25%	160.06
9.	$7,400.00	40%	2960	29.	$900.00	40%	360
10.	$658.75	35%	230.56	30.	$415.50	11⅜%	47.26
11.	$400.00	20%	80	31.	$21,750.00	22%	4785
12.	$925.50	15%	138.83	32.	$8,173.25	35%	2860.64
13.	$717.25	18%	129.11	33.	$767.20	16¾%	128.51
14.	$629.84	15%	94.48	34.	$2,480.25	45%	1116.11
15.	$3,985.15	10%	398.52	35.	$960.00	14.2%	136.32
16.	$200.00	12½%	25	36.	$817.20	20.8%	169.98
17.	$718.99	14%	100.66	37.	$415.10	16%	66.42
18.	$515.20	25%	128.80	38.	$9,000.00	9½%	855
19.	$7,600.00	40%	3040	39.	$11,400.00	7%	798
20.	$95,000.00	30%	28,500	40.	$880.15	6.8%	59.85

Page 36

1. $\frac{20}{21}$
2. $\frac{5}{8}$
3. $\frac{7}{30}$
4. $1\frac{7}{15}$
5. $\frac{9}{14}$
6. $1\frac{5}{22}$
7. $1\frac{6}{7}$
8. $1\frac{7}{15}$
9. $6\frac{2}{9}$
10. $\frac{12}{13}$
11. $\frac{4}{5}$
12. $2\frac{11}{12}$
13. $\frac{33}{40}$
14. $\frac{25}{36}$
15. $1\frac{7}{10}$
16. $9\frac{1}{3}$
17. $1\frac{1}{15}$
18. $\frac{32}{33}$
19. $1\frac{11}{70}$
20. $\frac{13}{28}$

Page 37

1. $1\frac{1}{16}$
2. $2\frac{5}{8}$
3. $3\frac{5}{24}$
4. $\frac{11}{18}$
5. $\frac{1}{2}$
6. $1\frac{1}{6}$
7. $\frac{24}{25}$
8. 2
9. $1\frac{1}{6}$
10. 11
11. $\frac{3}{8}$
12. $1\frac{4}{5}$
13. $1\frac{11}{45}$
14. $\frac{25}{32}$
15. $1\frac{1}{9}$
16. $5\frac{1}{2}$
17. $1\frac{23}{57}$
18. $\frac{3}{8}$
19. $2\frac{2}{5}$
20. $\frac{11}{36}$

Page 38

1. $\frac{40}{49}$
2. $2\frac{2}{11}$
3. $\frac{6}{25}$
4. $\frac{15}{16}$
5. $1\frac{1}{56}$
6. $1\frac{3}{5}$
7. $\frac{16}{25}$
8. $\frac{3}{16}$
9. $\frac{5}{22}$
10. 4
11. $\frac{1}{4}$
12. $3\frac{18}{23}$
13. $1\frac{1}{9}$
14. $1\frac{9}{16}$
15. $\frac{10}{27}$
16. $\frac{33}{35}$
17. $3\frac{3}{14}$
18. $\frac{27}{28}$
19. $\frac{3}{4}$
20. $9\frac{17}{22}$

Answer Key

Page 39
1. 8
2. 65
3. $1\frac{1}{4}$
4. 48
5. $4\frac{1}{2}$
6. $2\frac{1}{2}$
7. 105
8. $1\frac{1}{2}$
9. $3\frac{3}{7}$
10. 8
11. $11\frac{3}{7}$
12. 99
13. $22\frac{11}{12}$
14. $13\frac{1}{3}$
15. 18
16. $26\frac{2}{3}$
17. $14\frac{2}{3}$
18. $7\frac{3}{7}$
19. $20\frac{2}{9}$
20. $8\frac{8}{19}$

Page 40

Number		2	3	4	5	6	8	9	10
1.	12	X	X	X		X			
2.	18	X	X			X		X	
3.	52	X		X					
4.	65				X				
5.	76	X		X					
6.	90	X	X		X	X		X	X
7.	105		X		X				
8.	304	X		X			X		
9.	481								
10.	530	X			X				X
11.	720	X	X	X	X	X	X	X	X
12.	1,342	X							
13.	2,008	X		X			X		
14.	3,500	X		X	X				X
15.	5,896	X		X			X		
16.	1,485		X		X			X	
17.	3,744	X	X	X		X	X	X	
18.	51,840	X	X	X	X	X	X	X	X
19.	15,550	X			X				X
20.	62,937		X					X	
21.	32,768	X		X			X		
22.	59,049		X					X	
23.	31,250	X			X				X
24.	60,480	X	X	X	X	X	X	X	X
25.	98,415		X		X			X	

Page 41
1. 800,000
2. 9,000
3. 6,000,000
4. 4,000,000,000
5. 20,000,000
6. 5,000
7. 80,000,000
8. 15,000
9. 700,000
10. 300,000
11. 300,000
12. 30,000
13. 12,000,000
14. 120,000
15. 50,000
16. 20,000
17. 1,700
18. 2,300
19. 4,400
20. 40
21. 500
22. 250,000
23. 30,000
24. 40,000

Page 42
1. 2192 R1708
2. 8742
3. 7373
4. 11,109 R4203
5. 42,896
6. 134,862 R1633
7. 4622 R296
8. 93,606 R71
9. 9811
10. 16,451 R520
11. 5007 R1287
12. 6040
13. 13,037 R604
14. 99,003
15. 563,314
16. 3002 R677
17. 4553
18. 138,380 R5272
19. 71,426
20. 216,656 R1078
21. 7456
22. 114,983 R2429
23. 122,435 R194
24. 202,763 R3151

Page 43
1. 36,827 R4062
2. 47,762
3. 208,975
4. 505,847 R195
5. 9003
6. 2,889,784 R1852
7. 62,347
8. 7432
9. 266,752
10. 1,187,414 R1288
11. 44,444
12. 1,646,182 R3054
13. 98,751
14. 81,153
15. 31,765 R2069
16. 155,615 R857
17. 5438
18. 97,284 R313
19. 21,446
20. 50,734

Page 44
These multiplication problems are correct:
1, 4, 5, 6, 9, 10, 12, 13, 15, 16, 17, 18, 19, 20.

Page 45
These division problems are correct:
4, 5, 6, 7, 8, 10, 11, 13, 15, 16.

Answer Key

Page 46

For the number 3,024,598,136,670, the place value for:
1=hundred-thousands
2=ten-billions
3=ten-thousands and trillions
4=billions
5=hundred-millions
6=hundreds and thousands
7=tens
8=millions
9=ten-millions
0=ones and hundred-billions

The number 9,438,722,017,654 has:
2 ten-millions
4 ones
0 hundred-thousands
8 billions
3 ten-billions
7 thousands
6 hundreds
7 hundred-millions
9 trillions
5 tens
2 millions
4 hundred-billions
1 ten-thousand

The number is 6,120,857,439,834.

For the numbers 5,603,447,628,515 and 6,613,436,528,017, the place values that have the same numbers in each are: tens, thousands, ten-thousands, hundred-millions, billions, hundred-billions.

The number with the specified place values is 3,602,404,427,612.

Page 47

1. two and six hundred thousand, eight hundred thirty-six millionths
2. four thousand, six hundred sixty-nine and three thousand four hundred fifty-five ten-thousandths
3. one millionth
4. eleven and twenty-three thousand four hundred six hundred-thousandths
5. eight hundred seven and five thousandths
6. ninety-four thousand, one hundred three hundred-thousandths
7. six hundred eighty-two and fifty-nine thousandths
8. five and five thousand, five hundred fifty-five ten-thousandths
9. three thousand, six hundred three millionths
10. nine thousand, one hundred sixty-seven and twenty-two hundredths
11. 118.00497
12. 39,074.87
13. 5,011.009
14. 900.632
15. 0.890001
16. 0.040602
17. 14.068421
18. 50.1093
19. 0.00003
20. 0.33413
21. 9,056.0048
22. 9.0152
23. 542.688
24. 17.019
25. 1.871018

Page 48

1.–10.

11. 4
12. 8
13. 1
14. 7
15. 3
16. 2
17. 9
18. 0
19. 1
20. 0

21.–30. The number is: 8,160.395429.
31. The number is: 3,814.102512.

Page 49

1. >
2. >
3. <
4. <
5. <
6. <
7. <
8. =
9. <
10. >
11. <
12. >
13. =
14. >
15. <
16. <
17. >
18. >
19. <
20. >
21. 0.4689, 0.4698, 0.4869
22. 3.45427, 3.45572, 3.45725
23. 213.59992, 213.90552, 213.95002
24. 0.0066763, 0.0066766, 0.007606
25. 11.112222, 11.121121, 11.121211

Answer Key

26. 8.89988, 8.899881, 8.89998
27. 0.662731, 0.663726, 0.667231, 0.667321
28. 0.0074322, 0.0074342, 0.00744232, 0.0074482
29. 3,401,728.0036; 3,401,782.0036; 3,401,782.00361
30. 9.2888888, 9.882086, 9.8826802, 9.88286
31. 89.70065; 89.70007; 89.07653
32. 6.048809, 6.04009, 6.00498
33. 0.1123201, 0.112313, 0.1123103

34. 0.554554, 0.554545, 0.554455
35. 0.631139, 0.631124, 0.63098
36. 0.981, 0.98062, 0.980609
37. 862.0443201, 862.044302, 862.043402, 862.0044989
38. 0.0469, 0.04609, 0.004906, 0.004609
39. 10,881.188118; 10,880.188818; 10,880.188188; 10,818.188118
40. 47.770669, 47.77009, 47.707696, 47.707669

Page 50
1. 693.04268
2. 231.6058
3. 28.889455
4. 5,935.58719
5. 671.333105
6. 93.9186
7. 724.42298
8. 15.169718
9. 97.294603
10. 174.847
11. 797.18
12. 1,209.647
13. 23.8735
14. 175.3032
15. 6.708
16. 3,052.064204
17. 34.6788
18. 1.0
19. 1,429.095856
20. 13,658.757286
21. 1,270.879
22. 1.665817
23. 20.917
24. 1.0
25. 32.53305
26. 1,449.71356
27. 3567.176
28. 1.164797

Page 51
1. 3.6774
2. 0.00356
3. 558.5209
4. 35.0635
5. 2,864.662
6. 0.88165
7. 5.5682
8. 1.75
9. 1.010101
10. 2.79503
11. 0.487767
12. 6.08191
13. 1.87657
14. 0.33866
15. 908.9428
16. 242.90909
17. 127.47
18. 1.45808
19. 915.835
20. 4,871.825
21. 81.0664
22. 0.768787
23. 4.75769
24. 14.53577
25. 0.056769
26. 17.87652
27. 196.45679
28. 1,073.881
29. 6.891058
30. 0.08102
31. 0.210145
32. 4.89794
33. 1.88892
34. 18.639
35. 0.7784
36. 320.26567

Page 52
1. 370.723
2. 10
3. 50
4. 98.74473
5. 10.33679
6. 1.167791
7. 313.20666
8. 25.81288
9. 0.46657
10. 11.9957
11. 33.4437
12. 111.619791
13. 10,793.993254
14. 964.45174
15. 856.5
16. 94.31823
17. 967,832.007971
18. 1,593.54343
19. 81.6777
20. 38.8793
21. 93.999959
22. 4,782.684
23. 76.37379
24. 0.288881

Page 53

Number	x	1,000	0.01	100	10,000	0.001	10	0.1
1. 27.9		27,900	.279	2,790	279,000	.0279	279	2.79
2. 0.0618		61.8	.000618	6.18	618	.0000618	.618	.00618
3. 300.46		300,460	3.0046	30,046	3,004,600	.30046	3,004.6	30.046
4. 0.55		550	.0055	55	5,500	.00055	5.5	.055
5. 23.175		23,175	.23175	2,317.5	231,750	.023175	231.75	2.3175
6. 599.86		599,860	5.9986	59,986	5,998,600	.59986	5,998.6	59.986
7. 1.246		1,246	.01246	124.6	12,460	.001246	12.46	.1246
8. 0.008		8	.00008	.8	80	.000008	.08	.0008
9. 4,682.7		4,682,700	46.827	468,270	46,827,000	4.6827	46,827	468.27
10. 7.651		7,651	.07651	765.1	76,510	.007651	76.51	.7651
11. 0.00049		.49	.0000049	.049	4.9	.00000049	.0049	.000049
12. 86.3		86,300	.863	8,630	863,000	.0863	863	8.63
13. 9.72		9,720	.0972	972	97,200	.00972	97.2	.972
14. 0.8657		865.7	.008657	86.75	8,657	.0008657	8.657	.08657
15. 15,119.6		15,119,600	151.196	1,511,960	151,196,000	15.1196	151,196	1,511.96
16. 6.003		6,003	.06003	600.3	60,030	.006003	60.03	.6003
17. 577.78		577,780	5.7778	57,778	5,777,800	.57778	5,777.8	57.778
18. 1,064.89		1,064,890	10.6489	106,489	10,648,900	1.06489	10,648.9	106.489
19. 0.13246		132.46	.0013246	13.246	1,324.6	.00013246	1.3246	0.013246
20. 3.992		3,992	.03992	399.2	39,920	.003992	39.92	.3992
21. 12,981.4		12,981,400	129.814	1,298,140	129,814,000	12.9814	129,814	1,298.14
22. 0.0053		5.3	.000053	.53	53	.0000053	.053	.00053
23. 74.09		74,090	.7409	7,409	740,900	.07409	740.9	7.409
24. 6.0003		6000.3	.060003	600.03	60,003	.0060003	60.003	.60003
25. 9,086.5		9,086,500	90.865	908,650	90,865,000	9.0865	90,865	908.65

Answer Key

Page 54

1. 646.5732
2. 84.925
3. 0.47304
4. 248.911
5. 1.96251
6. 41.832
7. 3,423.648
8. 40.7946
9. 1.94485
10. 2,148.12
11. 1549.45
12. 63.9236
13. 0.343312
14. 0.496921
15. 0.361098
16. 2.9767
17. 571.38
18. 34.85612
19. 16.14096
20. 6.8328

Page 55

1. 0.000584
2. 0.05166
3. 0.008712
4. 0.076635
5. 0.031304
6. 0.006384
7. 0.071638
8. 0.000015
9. 0.0366
10. 0.026928
11. 0.048777
12. 0.004914
13. 0.0937728
14. 0.06312
15. 0.0086742
16. 0.00869
17. 0.003962
18. 0.035344
19. 0.095718
20. 0.000507
21. 0.050282
22. 0.068476
23. 0.062976
24. 0.0876897

Page 56

Number	+	0.01	10,000	100	0.0001	0.1	1,000	0.001
1. 63.8		6,380	.00638	.638	638,000	638	.0638	63,800
2. 0.0092		.92	.00000092	.000092	92	.092	.0000092	9.2
3. 718.4		71,840	.07184	7.184	7,184,000	7,184	.7184	718,400
4. 9.663		966.3	.0009663	.09663	96,630	96.63	.009663	9,663
5. 500.6		50,060	.05006	5.006	5,006,000	5,006	.5006	500,600
6. 0.00081		.081	.000000081	.0000081	8.1	.0081	.00000081	.81
7. 8.005		800.5	.0008005	.08005	80,050	80.05	.008005	8,005
8. 40.067		4,006.7	.0040067	.40067	400,670	400.67	.040067	40,067
9. 7,100.5		710,050	.71005	71.005	71,005,000	71,005	7.1005	7,100,500
10. 0.06123		6.123	.000006123	.0006123	612.3	.6123	.00006123	61.23
11. 19.63		1,963	.001963	.1963	196,300	196.3	.01963	19,630
12. 44.441		4,444.1	.0044441	.44441	444,410	444.41	.044441	44,441
13. 1,003.6		100,360	.10036	10.036	10,036,000	10,036	1.0036	1,003,600
14. 6.022		602.2	.0006022	.06022	60,220	60.22	.006022	6,022
15. 0.00055		.055	.000000055	.0000055	5.5	.0055	.00000055	.55
16. 21,560.3		2,156,030	2.15603	215.603	215,603,000	215,603	21.5603	21,560,300
17. 0.1399		13.99	.00001399	.001399	1,399	1.399	.0001399	139.9
18. 20.441		2,044.1	.0020441	.20441	204,410	204.41	.020441	20,441
19. 7,115.8		711,580	.71158	71.158	71,158,000	71,158	7.1158	7,115,800
20. 6.8897		688.97	.00068897	.068897	68,897	68.897	.0068897	6,889.7
21. 17.099		1,709.9	.0017099	.17099	170,990	170.99	.017099	17,099
22. 321.05		32,105	.032105	3.2105	3,210,500	3,210.5	.32105	321,050
23. 69.4003		6,940.03	.00694003	.694003	694,003	694.003	.0694003	69,400.3
24. 1.0008		100.08	.00010008	.010008	10,008	10.008	.0010008	1,000.8
25. 555.11		55,511	.055511	5.5511	5,551,100	5,551.1	.55511	555,110

Page 57

1. 0.0745
2. 68.2
3. 5.881
4. 7.02
5. 40.7
6. 711.26
7. 0.542
8. 3.441
9. 0.00022
10. 0.203
11. 6.55
12. 1.008
13. 0.2007
14. 9.87
15. 0.063
16. 7.07
17. 34.7
18. 0.0999
19. 77.21
20. 0.1673
21. 1.698
22. 222.2
23. 4.77
24. 3.0047

Page 58

1. 78.5
2. 8.007
3. 0.0005
4. 84.03
5. 0.9007
6. 0.0211
7. 1.266
8. 9.44
9. 0.0077
10. 90.3
11. 9.001
12. 0.044
13. 0.554
14. 90.5
15. 1.88
16. 5.55
17. 7.99
18. 0.05
19. 0.007
20. 75.6
21. 0.0093
22. 0.344
23. 5.64
24. 31.022

Page 59

	Measurement	Precision to the Nearest	GPE	Actual Length
1.	82 hm	hm	0.5 hm	82 hm ± 0.5 hm
2.	9 dg	decigram	0.5 dg	9 dg ± 0.5 dg
3.	247 L	liter	0.5 L	247 L ± 0.5 L
4.	35 cm	centimeter	0.5 cm	35 cm ± 0.5 cm
5.	49 kg	Kg	0.5 kg	49 kg ± 0.5 kg
6.	112 mL	milliliter	0.5 mL	112 mL ± 0.5 mL
7.	6 dam	dekameter	.5 dam	6 dam ± 0.5 dam
8.	75 kg	kilogram	0.5 kg	75 kg ± 0.5 kg
9.	86 cL	cL	0.5 cL	86 cL ± 0.5 cL
10.	14 hm	hectometer	0.5 hm	14 hm ± 0.5 hm
11.	647 mm	millimeter	0.5 mm	647 mm ± 0.5 mm
12.	51 mg	milligram	0.5 mg	51 mg ± 0.5 mg
13.	33 dL	dL	0.5 dL	33 dL ± 0.5 dL
14.	240 km	kilometer	0.5 km	240 km ± 0.5 km
15.	467 cg	centigram	0.5 cg	467 cg ± 0.5 cg
16.	21.3 cm	0.1 centimeter	0.05 cm	21.3 cm ± 0.05 cm
17.	346.09 L	.01 L	0.005 L	346.09 L ± 0.005 L
18.	3.7 cm	0.1 centimeter	0.05 cm	3.7 cm ± 0.05 cm
19.	29.88 g	0.01 gram	0.005 g	29.88 g ± 0.005 g
20.	9.2 hL	0.1 hectoliter	0.05 hL	9.2 hL ± 0.05 hL
21.	10.17 dam	.01 dam	0.005 dam	10.17 dam ± 0.005 dam
22.	200 mm	millimeter	0.5 mm	200 mm ± 0.5 mm
23.	918.01 cm	0.01 centimeter	0.005 cm	918.01 cm ± 0.005 cm
24.	63.9 L	0.1 liter	0.05 L	63.9 L ± 0.05 L
25.	4.003 g	.001 g	0.0005 g	4.003 g ± 0.0005 g
26.	26.4 kg	0.1 kilogram	0.05 kg	26.4 kg ± 0.05 kg
27.	30.7 m	0.1 meter	0.05 m	30.7 m ± 0.05 m
28.	16.42 kg	0.01 kilogram	0.005 kg	16.42 kg ± 0.005 kg
29.	1.111 kL	.001 kL	0.0005 kL	1.111 kL ± 0.0005 kL
30.	114.14 dm	0.01 decimeter	0.005 dm	114.14 dm ± 0.005 dm

Answer Key

© Carson-Dellosa • CD-704389

Page 60
1. prime
2. 2, 15, 3, 10, 5, 6
3. 3, 17
4. 5, 11
5. 2, 30, 3, 20, 4, 15, 5, 12, 6, 10
6. 5
7. 5, 15, 3, 25
8. 2, 9, 3, 6
9. prime
10. 2, 8, 4
11. prime
12. 3, 19
13. 2, 50, 4, 25, 5, 20, 10
14. 2, 6, 3, 4
15. 2, 12, 3, 8, 4, 6
16. 2, 14, 4, 7
17. 2, 18, 3, 12, 4, 9, 6
18. prime
19. 7
20. 7, 11
21. 2, 40, 4, 20, 5, 16, 8, 10
22. prime
23. 2, 32, 4, 16, 8
24. 3, 5
25. 5, 13
26. 2, 25, 5, 10
27. prime
28. 2, 7
29. prime
30. prime

Page 61
1. $2^6 \times 5$
2. $2^2 \times 5^3$
3. $3^3 \times 2^4$
4. $2^3 \times 13$
5. $2^5 \times 11$
6. $3^4 \times 19$
7. $2^3 \times 5^3$
8. $2^2 \times 7^3$
9. $2^5 \times 71$
10. $2^3 \times 3^2 \times 11$
11. $2 \times 3 \times 11 \times 13$
12. $2^2 \times 3 \times 5 \times 17$
13. $3^2 \times 5^3$
14. $2^4 \times 7^2 \times 11$
15. $2 \times 3 \times 5 \times 7 \times 11 \times 13$
16. $2^2 \times 3 \times 11 \times 23$
17. $2^2 \times 3^2 \times 5^2$
18. 11×19^2
19. $3^3 \times 5^3$
20. $2^2 \times 3^2 \times 11 \times 17$
21. $2^3 \times 37$
22. $5 \times 7 \times 41$
23. $2^6 \times 7 \times 13$
24. $2^2 \times 3^2 \times 17^2$
25. $7^2 \times 11^2$
26. $2 \times 3 \times 5 \times 7^2 \times 11$
27. $3^5 \times 7 \times 11$
28. $2 \times 5^2 \times 7^4$

Page 62
1. 20
2. 21
3. 15
4. 8
5. 90
6. 21
7. 4
8. 4
9. 3
10. 2
11. 5
12. 10
13. 2
14. 2
15. 3
16. 1
17. 4
18. 1
19. 2
20. 9
21. 9
22. 7
23. 9
24. 5
25. 1
26. 4
27. 6
28. 10

Page 63
1. 18
2. 14
3. 24
4. 20
5. 18
6. 9
7. 45
8. 6
9. 18
10. 30
11. 30
12. 84
13. 60
14. 12
15. 28
16. 24
17. 36
18. 28
19. 12
20. 90
21. 66
22. 12
23. 42
24. 12
25. 39
26. 18

Page 64
1. 5, 225
2. 17, 255
3. 3, 126
4. 2, 936
5. 2, 348
6. 5, 195
7. 22, 132
8. 4, 312
9. 2, 840
10. 4; 1,428
11. 10, 420
12. 4, 448
13. 7, 385
14. 5, 220
15. 15, 300
16. 8, 96
17. 11, 132
18. 2, 552
19. 19, 228
20. 28, 168
21. 4, 80
22. 11, 110
23. 5, 650
24. 11, 462
25. 10, 150
26. 9, 630
27. 4, 560
28. 9, 162
29. 19, 114
30. 12, 144
31. 5, 350
32. 7, 196

Page 65
1. -2° C
2. -3° C
3. 25° C
4. 80° C
5. -76
6. 14
7. 6
8. -25
9. 34
10. 37
11. -14
12. -78
13. 57
14. -29
15. 15
16. -5

Answer Key

Page 66

1. negative
2. positive
3. positive
4. negative
5. negative
6. positive
7. positive
8. negative
9. negative
10. negative
11. positive
12. positive
13. negative
14. negative
15. positive
16. -42
17. 7
18. 12
19. 15
20. -21
21. -106
22. 230
23. 81
24. 60
25. -75

26. 111
27. -525
28. 65
29. 33
30. 2
31. -6
32. 1
33. 5
34. -9
35. -4
36. 4
37. -10
38. 3
39. 10
40. -1
41. 9
42. -7
43. -3
44. 6
45. 2
46. -5
47. 7
48. -8
49. 8
50. -2

Page 67

1. $\dfrac{6}{10}$
2. $\dfrac{25}{7}$
3. $\dfrac{-10}{1}$
4. $\dfrac{-82}{100}$
5. $\dfrac{-333}{100}$
6. $\dfrac{-35}{6}$
7. $\dfrac{212}{100}$
8. $\dfrac{85}{100}$
9. $\dfrac{27}{1}$

10. $\dfrac{-68}{100}$
11. $\dfrac{-9}{1}$
12. $\dfrac{-836}{100}$
13. $\dfrac{-487}{100}$
14. $\dfrac{-60}{7}$
15. $\dfrac{25}{2}$
16. $\dfrac{44}{100}$
17. $\dfrac{-16}{100}$
18. $\dfrac{103}{100}$

19. $-\dfrac{299}{100}$
20. $\dfrac{-24}{100}$
21. $\dfrac{525}{100}$
22. $\dfrac{72}{1}$
23. $\dfrac{73}{10}$
24. $\dfrac{28}{100}$
25. $\dfrac{-32}{5}$
26. >
27. <
28. >
29. =
30. <
31. >
32. =

Page 68

1. 11
2. 28
3. 33
4. 110
5. 50
6. 35
7. 4
8. 18
9. 72
10. 18
11. 25
12. 71
13. 64
14. 44

33. >
34. >
35. <
36. =
37. <
38. =
39. >
40. >
41. -12.74, 12.73, $12\dfrac{11}{15}$, $12\dfrac{3}{4}$
42. $-4\dfrac{2}{9}$, -4.201, $-4\dfrac{1}{5}$, -4.19
43. $\dfrac{6}{25}$, 0.252, $\dfrac{4}{15}$, $\dfrac{6}{25}$
44. -1.401, $-1\dfrac{2}{5}$, $-1\dfrac{3}{8}$, 1.389
45. $15\dfrac{3}{20}$, 15.151, 15.185, $15\dfrac{3}{16}$

15. 36
16. 41
17. 8
18. 9
19. 214
20. 510
21. 40
22. 51
23. 60
24. 21
25. 4
26. 42
27. 25
28. 22

Page 69

1. 23
2. 5
3. 20
4. 54
5. 5
6. 11
7. 40
8. 19

9. 6
10. 23
11. 64
12. 10
13. 14
14. 30
15. 9
16. 6

Answer Key

17. 5
18. 8
19. 10
20. 100

Page 70
1. Distributive
2. Identity
3. Commutative
4. Associative
5. Identity
6. Distributive
7. Associative
8. Commutative
9. 0
10. 9

21. 3
22. 28
23. 1,296
24. 5

11. -11
12. -5
13. 14
14. -55
15. -13
16. -15
17. 1
18. -75
19. 4
20. -9

Page 71
1. 29
2. -24
3. -19
4. -14
5. -24
6. 21
7. -20
8. -12
9. -22
10. -16
11. -21
12. -21
13. -13

14. -39
15. -33
16. 60
17. -22
18. -55
19. -41
20. -37
21. -36
22. -88
23. -21
24. 97
25. -23
26. 40

Page 72
1. -8
2. 16
3. -7
4. -3
5. 6
6. 48
7. 15
8. -62
9. 58
10. -52
11. -4
12. -18
13. -42
14. -15
15. 23
16. -25

17. 17
18. 9
19. -12
20. 12
21. 22
22. -16
23. -25
24. 85
25. 57
26. 27
27. -108
28. -41
29. -43
30. 69
31. 0
32. 313

Page 73
1. -112
2. 30
3. -52
4. 107
5. 18
6. -83
7. 122
8. -29
9. -21
10. -186
11. -55
12. -540
13. 20

14. -70
15. 41
16. -55
17. 98
18. -100
19. 33
20. -67
21. 30
22. -45
23. -22
24. 178
25. 93
26. -12

Page 74
1. -42
2. 20
3. 265
4. 3
5. 214
6. -19

7. -411
8. 4
9. 36
10. -88
11. 144
12. 12

Page 75
1. 208
2. -95
3. 72
4. -62
5. 1
6. 125

7. -117
8. 106
9. -143
10. 313
11. -20
12. 244

Page 76
1. 63,224,700
2. 6,197,488
3. 111,276
4. -151,188
5. -72,998
6. -3,990602
7. -3,669,660
8. 17,770,384
9. 208,145
10. -102,604

11. -1,148,344
12. 25,572,250
13. 1,869,395
14. 20,124
15. -2,157,453
16. 14,902,096
17. -2,558,010
18. 559,152
19. -168,412
20. 845,432

Answer Key

Page 77
1. -516
2. 911
3. 119
4. -608
5. 193
6. -284
7. -144
8. -1,112
9. -306
10. 514
11. 409
12. 82
13. 228
14. -97
15. -813
16. -68
17. -108
18. 648
19. 983
20. 609

Page 78
1. 20
2. -88
3. -64
4. 10
5. -42
6. -8
7. 528
8. 20
9. -150
10. 54
11. -45
12. -36
13. 3
14. 15
15. -21

Page 79
1. $0.\overline{18}$
2. $0.8\overline{3}$
3. $3.5\overline{3}$
4. $2.07\overline{4}$
5. $13.6\overline{1}$
6. $1.13\overline{6}$
7. $0.\overline{15}$
8. $0.38\overline{63}$
9. $0.0208\overline{3}$
10. $6.3\overline{6}$
11. $0.\overline{428571}$
12. $0.7\overline{6}$
13. $0.458\overline{3}$
14. $2.1\overline{3}$
15. $0.3\overline{6}$
16. $0.\overline{7}$
17. $0.\overline{054}$
18. $2.\overline{5}$
19. $108.\overline{6}$
20. $4.02\overline{7}$
21. $4.\overline{6}$
22. $14.1\overline{6}$
23. $5.\overline{51}$

Page 80
1. $0.01\overline{8}$
2. $0.291\overline{6}$
3. $0.1\overline{5}$
4. $0.9\overline{4}$
5. $0.97\overline{2}$
6. $218.8\overline{3}$
7. $0.0\overline{75}$
8. $1.\overline{01}$
9. $54.7\overline{2}$
10. $0.09\overline{25}$
11. $0.\overline{84}$
12. $2.\overline{692307}$
13. $3.4\overline{07}$
14. $1.11\overline{6}$
15. $2.1\overline{5}$
16. $12.68\overline{1}$
17. $2.2\overline{3}$
18. $0.2\overline{7}$
19. $29.\overline{2}$
20. $25.\overline{3}$

Page 81
1. (5 × 5 array of dots)
2. 25
3. 343
4. 64
5. 512
6. 81
7. 625
8. 27
9. 32
10. 216
11. 8
12. 23
13. 33
14. 100
15. 8
16. 765
17. 22
18. 37
19. 17
20. $41\frac{2}{3}$
21. 4
22. 20
23. 16
24. 40

Page 82
1. $\dfrac{25}{36}$
2. $\dfrac{49}{100}$
3. $\dfrac{27}{64}$
4. $\dfrac{81}{169}$
5. $\dfrac{1}{121}$
6. $\dfrac{16}{225}$
7. $\dfrac{1}{16}$
8. $\dfrac{1}{27}$
9. $\dfrac{25}{49}$
10. $\dfrac{8}{125}$
11. $\dfrac{9}{64}$
12. $\dfrac{49}{81}$
13. $\dfrac{9}{625}$
14. $\dfrac{1}{216}$
15. $\dfrac{25}{256}$
16. $\dfrac{16}{49}$
17. $\dfrac{49}{64}$
18. $\dfrac{25}{144}$
19. $\dfrac{64}{125}$
20. $\dfrac{121}{400}$

Page 83
1. 22
2. 32
3. 10
4. 75
5. 9
6. 20
7. 8
8. 6
9. 99.5
10. 170
11. 17
12. 46

Answer Key

13. 96
14. 34
15. 60
16. 7

Page 84
1. 7
2. 13
3. 28
4. 15
5. 8
6. 20
7. 25
8. 26
9. 43

17. 52
18. 2
19. 156
20. 55

10. 50
11. 70
12. 50
13. 10
14. 70
15. 70
16. 48.4
17. 45
18. 1,500

Page 85
1. r − 4
2. $\dfrac{y}{2}$
3. p + 7
4. d3
5. $\dfrac{12}{g}$
6. k − 5
7. y + 5
8. 16 − r
9. $\dfrac{10}{p}$

10. 6z
11. $\dfrac{m}{3}$
12. b + 4
13. t times 4
14. 5 plus h
15. r divided by 5
16. w times 6
17. q plus 2
18. 44 divided by f
19. 4.9, 68, .172
20. .555, 52.535, 2.835

Page 86
1. 24.471
2. 42.8
3. 2.1
4. 63.9
5. 469,238
6. 2.26
7. s = 30, m = 4, n = 2, j = 8, k = 3

8. 11.17
9. 4
10. $1\dfrac{2}{3}$
11. a = 6, b = 5, t = 2
12. c = 2, m = 4, p = 8, r = 7, s = 10

Page 87
1. 14
2. 2
3. 14
4. 3
5. 3
6. 3
7. 20
8. 2

9. 14
10. 36
11. 12
12. 25
13. 28
14. 3
15. 4
16. 29

17. 5
18. 41
19. 2
20. 9
21. -12
22. 16
23. -10
24. 49
25. 61
26. 15
27. 4
28. -30

29. 27
30. -140
31. 8
32. 9
33. 100
34. 50
35. 6
36. 2
37. 7
38. 9
39. 25
40. 5

Page 88
1. no
2. no
3. yes
4. yes
5. no
6. yes
7. yes
8. no
9. yes
10. no

11. yes
12. no
13. no
14. no
15. yes
16. no
17. yes
18. no
19. yes
20. yes

Page 89
1. 3
2. 30
3. 9
4. 13
5. -17
6. 13
7. 3
8. -13
9. 9
10. 15
11. 5
12. 5
13. 24
14. -8
15. 12
16. 8
17. -25
18. 31
19. -21
20. -21

21. -20
22. -12
23. 2
24. 11
25. 11
26. -11
27. -12
28. 20
29. -23
30. -31
31. -25
32. 5
33. 25
34. -38
35. 6
36. 37
37. -21
38. 70
39. 3
40. 6

Answer Key

Page 90
1. -5
2. 18
3. 8
4. -44
5. -28
6. -7
7. -40
8. 9
9. 42
10. -15
11. 99
12. -16
13. 4
14. -72
15. -11
16. 65
17. -15
18. 72
19. 6
20. -130
21. -88
22. 4
23. 135
24. 10
25. 13
26. 40
27. 48
28. -45
29. 9
30. -200
31. -11
32. 81
33. 84
34. -3
35. 7
36. -80
37. 9
38. 84
39. 8
40. -26

Page 91
1. 7
2. 32
3. -8
4. 10
5. 2
6. 55
7. 8
8. 4
9. 4
10. -8
11. 2
12. 3
13. 4
14. 5
15. 3
16. 63
17. 12
18. 48
19. 7
20. 3
21. 3
22. 20
23. 10
24. 77
25. 8
26. -44
27. 75
28. 5
29. -24
30. 7
31. 110
32. 8
33. 5
34. -60
35. 4
36. -75

Page 92
1. 9 square units
2. 8 square units
3. $8\frac{1}{2}$ square units
4. 10 square units
5. 2.5 square units
6. 12 square units
7. 9 square units
8. 9 square units
9. 6 square units
10. 5 square units
11. 13 square units
12. 9 square units

Page 93
1. 16 square units
2. 8 square units
3. 10 square units
4. Shapes will vary.
5. Shapes will vary.
6. Shapes will vary.
7. Shapes will vary.
8. Shapes will vary.
9. Shapes will vary.

Page 94
1. 52 square inches
2. 2,500 square inches
3. 8,019 square inches
4. 608 square inches
5. 152 square inches
6. 52 square inches
7. 110 square inches
8. 45 square inches
9. 12 square centimeters
10. $24\frac{1}{2}$ square centimeters
11. 24 square centimeters

Page 95
1. 40 square centimeters
2. 84 square centimeters
3. 12 square centimeters
4. 60 square inches
5. 64 square inches
6. 80 square inches
7. 18 square inches
8. 12 square inches
9. 40 square inches
10. 5 square inches

Page 96
1. 202 square inches
2. 282 square millimeters
3. 104 square centimeters
4. 216 square meters
5. 210 square inches
6. 54 square feet
7. 106 square centimeters
8. 90 square meters
9. 104 square meters
10. 960 square feet
11. 158 square inches
12. 256 square millimeters

Answer Key

Page 97

1. 100.32 square units
2. 113.04 square units
3. 1,188 square units
4. 144 square units
5. 3.84 square units
6. 5.76 square units
7. 200π square units
8. 36π square units
9. 6π square units

Page 98

1. 108 cubic inches
2. 60 cubic inches
3. 35 cubic inches
4. 72 cubic inches
5. 32 cubic inches
6. 96 cubic inches
7. 54 cubic inches
8. 27 cubic inches
9. 30 cubic centimeters
10. 12 cubic centimeters
11. 28 cubic meters

Page 99

1. 54 cubic feet
2. 29,440 cubic millimeters
3. 432 cubic yards
4. 196 cubic centimeters
5. 9,000 cubic meters
6. 2,464 cubic centimeters
7. 105 cubic inches
8. 1,794 cubic meters
9. 96 cubic meters
10. 4,096 cubic centimeters
11. 1,728
12. 27
13. 1.5 or $1\frac{1}{2}$
14. 3
15. 46,656
16. 3

Page 100

1. $\overline{AH}$, CG, DE
2. DE
3. I
4. CI, BI, FI, GI
5. CG
6. FG
7. ∠FIG, ∠CIF, ∠CIB
8. 128.74 mm
9. 116.18 mm
10. 87.92 mm

Page 101

1. 15.7 in.
2. 12.56 in.
3. 9.42 in.
4. 78.5 square inches
5. 153.86 square inches
6. 28.26 square inches
7. 346.185 square millimeters
8. 615.44 square millimeters
9. 1,256 square millimeters

Page 102

1–5. Acute angles are AOB, AOC, BOC, BOD, COD, DOE, EOF. Obtuse angles are AOE, BOE, BOF, COE, COF. Right angles are AOD, DOF.

6. supplementary
7. supplementary
8. vertical
9. vertical
10. corresponding
11. corresponding
12. vertical
13. supplementary
14. 45°
15. 25°
16. 95°

Page 103

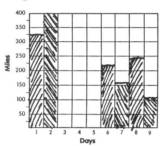

1. 1,453
2. 220
3. Day 1 (325), Day 2 (400)
4. They probably visited a single place.
5. 50:80
6. 70:95
7. 55:100
8. 115:195
9. 105:180
10. 145:275

Page 104

1. 20:160
2. 105:250
3. 88:120
4. 135:290
5. 36:130
6. 140:410
7. 165
8. 170, 172, 172, 173, 174, 174, 175, 177
9. 165
10. 166
11. 158

Page 105

Set	Mean	Mode	Median	Range
1	52.4	35	41	48
2	12	12	12	14
3	116	132	115	34

Set	Mean	Mode	Median	Range
4	6	5	5	5
5	44.3	NONE	46	37
6	37.7	4	16	97

7. 25
8. 37
9. 38
10. 31.75
11. 13
12. 16
13. 19
14. 14

Answer Key

Page 106

	range	median	mean	mode
1.	38	96	103	90
2.	272	476	458	501
3.	425	1,315	1,277	1,442
4.	75	57	52	57
5.	9	95	95	98
6.	150	347	308	367
7.	4,947	55,866	55,237	55,866
8.	12	5	5	3
9.	999	100,101	100,152	100,110
10.	24	2,033	2,029	2,035

Page 107

	range	median	mean	mode
1.	396	462,151	462,295	462,511
2.	1,014	51,105	50,664	51,110
3.	16	1,616	1,615	1,621
4.	14	310	309	310
5.	3,644	44,000	42,420	44,004
6.	8	567	566	568
7.	33	33	30	19
8.	704	771,117	771,232	771,711

Page 108

1. 1:8
2. 2:8
3. 2:8
4. 2:8
5. 3:8
6. 0:8

7. 12:24
8. 6:24
9. 8:24
10. 4:24
11. 2:24
12. 1:24

Page 109

1. 29:200
2. 139:410
3. 41:190
4. 93:400
5. 95:210
6. 232:600
7. 1:8
8. 6:8
9. 6:8
10. 8:8
11. 0:8
12. 5:8

Page 110

1. 1:6
2. 5:6
3. 3:6
4. 5:6
5. 1:6
6. 4:6
7. 3:6
8. 5:6
9. 2:6
10. 3:6
11. 1:8
12. 3:8
13. 2:8
14. 4:8
15. 2:8
16. 5:8
17. 3:8
18. 3:8
19. 7:8
20. 6:8